A
Mathematician
Plays the
Stock Market

Also by John Allen Paulos

Mathematics and Humor (1980)

I Think Therefore I Laugh (1985)

*Innumeracy: Mathematical Illiteracy
and its Consequences (1988)*

*Beyond Numeracy: Ruminations
of a Numbers Man (1991)*

*A Mathematician Reads
the Newspaper (1995)*

*Once Upon a Number: The Hidden
Mathematical Logic of Stories (1998)*

A
Mathematician
Plays the
Stock Market

John Allen Paulos

BASIC

BOOKS

OCM 55687117

A Member of the Perseus Books Group

New York

Designed by Trish Wilkinson
Set in 11-point Sabon by the Perseus Books Group

The Library of Congress cataloged the hardcover edition as follows:
Paulos, John Allen.
    A mathematician plays the stock market / John Allen Paulos.
        p.   cm.
    Includes bibliographical references and index.
    ISBN 0-465-05480-3 (hc.)
    1. Investments—Psychological aspects. 2. Stock
exchanges—Psychological aspects. 3. Stock exchanges—Mathematical
models. 4. Investment analysis. 5. Stocks. I. Title.
    ISBN 0-465-05481-1 (pbk.)

HG4515.15.P38 2003
332.63'2042—dc21   2002156215

04 05 06 / 10 9 8 7 6 5 4 3 2 1

*To my father, who never played
the market and knew little about probability,
yet understood one of the prime lessons of both.
"Uncertainty," he would say, "is the only
certainty there is, and knowing how to live
with insecurity is the only security."*

Contents

1 | Anticipating Others' Anticipations

*I*t was early 2000, the market was booming, and my investments in various index funds were doing well but not generating much excitement. Why investments should generate excitement is another issue, but it seemed that many people were genuinely enjoying the active management of their portfolios. So when I received a small and totally unexpected chunk of money, I placed it into what Richard Thaler, a behavioral economist I'll return to later, calls a separate mental account. I considered it, in effect, "mad money."

Nothing distinguished the money from other assets of mine except this private designation, but being so classified made my modest windfall more vulnerable to whim. In this case it entrained a series of ill-fated investment decisions that, even now, are excruciating to recall. The psychological ease with which such funds tend to be spent was no doubt a factor in my using the unexpected money to buy some shares of WorldCom (abbreviated WCOM), "the pre-eminent global communications company for the digital generation," as its ads boasted, at $47 per share. (Hereafter I'll generally use WCOM to refer to the stock and WorldCom to refer to the company.)

Today, of course, WorldCom is synonymous with business fraud, but in the halcyon late 1990s it seemed an irrepressibly

successful devourer of high-tech telecommunications companies. Bernie Ebbers, the founder and former CEO, is now
viewed by many as a pirate, but then he was seen as a swash-
buckler. I had read about the company, knew that high-tech
guru George Gilder had been long and fervently singing its
praises, and was aware that among its holdings were MCI,
the huge long-distance telephone company, and UUNet, the
"backbone" of the Internet. I spend a lot of time on the net
(home is where you hang your @) so I found Gilder's lyrical
writings on the "telecosm" and the glories of unlimited bandwidth particularly seductive.

I also knew that, unlike most dot-com companies with no
money coming in and few customers, WorldCom had more
than $25 billion in revenues and almost 25 million customers,
and so when several people I knew told me that WorldCom
was a "strong buy," I was receptive to their suggestion. Although the stock had recently fallen a little in price, it was, I
was assured, likely to soon surpass its previous high of $64.

If this was all there was to it, there would have been no important financial consequences for me, and I wouldn't be writing about the investment now. Alas, there was something else,
or rather a whole series of "something elses." After buying the
shares, I found myself idly wondering, why not buy more? I
don't think of myself as a gambler, but I willed myself *not* to
think, willed myself simply to act, willed myself to buy more
shares of WCOM, shares that cost considerably more than the
few I'd already bought. Nor were these the last shares I would
buy. Usually a hardheaded fellow, I was nevertheless falling
disastrously in love.

Although my particular heartthrob was WCOM, almost all
of what I will say about my experience is unfortunately appli-
cable to many other stocks and many other investors. Wher-
ever WCOM appears, you may wish to substitute the symbols

for Lucent, Tyco, Intel, Yahoo, AOL-Time Warner, Global Crossing, Enron, Adelphia, or, perhaps, the generic symbols WOE or BANE. The time frame of the book—in the midst of a market collapse after a heady, nearly decade-long surge—may also appear rather more specific and constraining than it is. Almost all the points made herein are rather general or can be generalized with a little common sense.

Falling in Love with WorldCom

John Maynard Keynes, arguably the greatest economist of the twentieth century, likened the position of short-term investors in a stock market to that of readers in a newspaper beauty contest (popular in his day). The ostensible task of the readers is to pick the five prettiest out of, say, one hundred contestants, but their real job is more complicated. The reason is that the newspaper rewards them with small prizes only if they pick the five contestants who receive the most votes from readers. That is, they must pick the contestants that they think are most likely to be picked by the other readers, and the other readers must try to do the same. They're not to become enamored of any of the contestants or otherwise give undue weight to their own taste. Rather they must, in Keynes' words, anticipate "what average opinion expects the average opinion to be" (or, worse, anticipate what the average opinion expects the average opinion expects the average opinion to be).

Thus it may be that, as in politics, the golden touch derives oddly from being in tune with the brass masses. People might dismiss rumors, for example, about "Enronitis" or "World-Comism" affecting the companies in which they've invested, but if they believe others will believe the rumors, they can't afford to ignore them.

BWC (before WorldCom) such social calculations never interested me much. I didn't find the market particularly inspiring or exalted and viewed it simply as a way to trade shares in businesses. Studying the market wasn't nearly as engaging as doing mathematics or philosophy or watching the Comedy Network. Thus, taking Keynes literally and not having much confidence in my judgment of popular taste, I refrained from investing in individual stocks. In addition, I believed that stock movements were entirely random and that trying to outsmart dice was a fool's errand. The bulk of my money therefore went into broad-gauge stock index funds.

AWC, however, I deviated from this generally wise course. Fathoming the market, to the extent possible, and predicting it, if at all possible, suddenly became live issues. Instead of snidely dismissing the business talk shows' vapid talk, sportscaster-ish attitudes, and empty prognostication, I began to search for what of substance might underlie all the commentary about the market and slowly changed my mind about some matters. I also sought to account for my own sometimes foolish behavior, instances of which will appear throughout the book, and tried to reconcile it with my understanding of the mathematics underlying the market.

Lest you dread a cloyingly personal account of how I lost my shirt (or at least had my sleeves shortened), I should stress that my primary purpose here is to lay out, elucidate, and explore the basic conceptual mathematics of the market. I'll examine—largely via vignettes and stories rather than formulas and equations—various approaches to investing as well as a number of problems, paradoxes, and puzzles, some old, some new, that encapsulate issues associated with the market. Is it efficient? Random? Is there anything to technical analysis, fundamental analysis? How can one quantify risk? What is the role of cognitive illusion? Of common knowledge? What are the most common scams? What are options, portfolio theory, short-selling, the efficient market hypothesis? Does the normal bell-shaped curve explain the market's occasional extreme volatility? What about fractals, chaos, and other non-standard tools? There will be no explicit investment advice and certainly no segments devoted to the ten best stocks for the new millennium, the five smartest ways to jump-start your 401(k), or the three savviest steps you can take right now. In short, there'll be no financial pornography.

Often inseparable from these mathematical issues, however, is psychology, and so I'll begin with a discussion of the no-man's land between this discipline and mathematics.

Being Right Versus
Being Right About the Market

There's something very reductive about the stock market. You can be right for the wrong reasons or wrong for the right reasons, but to the market you're just plain right or wrong. Compare this to the story of the teacher who asks if anyone in the class can name two pronouns. When no one volunteers, the teacher calls on Tommy who responds, "Who, me?" To the market, Tommy is right and therefore, despite being unlikely to get an A in English, he's rich.

Guessing right about the market usually leads to chortling. While waiting to give a radio interview at a studio in Philadelphia in June 2002, I mentioned to the security guard that I was writing this book. This set him off on a long disquisition on the market and how a couple of years before he had received two consecutive statements from his 401(k) administrator indicating that his retirement funds had declined. (He took this to be what in chapter 3 is called a technical sell signal.) "The first one I might think was an accident, but two in a row, no. Do you know I had to argue with that pension person there about getting out of stocks and into those treasury bills? She told me not to worry because I wasn't going to retire for years, but I insisted 'No, I want out now.' And I'm sure glad I did get out." He went on to tell me about "all the big shots at the station who cry like babies every day about how much money they lost. I warned them that two down statements and you get out, but they didn't listen to me."

I didn't tell the guard about my ill-starred WorldCom experience, but later I did say to the producer and sound man that the guard had told me about his financial foresight in response to my mentioning my book on the stock market. They both assured me that he would have told me no matter what. "He tells everyone," they said, with the glum humor of big shots who didn't take his advice and now cry like babies.

Such anecdotes bring up the question: "If you're so smart, why ain't you rich?" Anyone with a modicum of intelligence and an unpaid bill or two is asked this question repeatedly. But just as there is a distinction between being smart and being rich, there is a parallel distinction between being right and being right about the market.

Consider a situation in which the individuals in a group must simultaneously choose a number between 0 and 100. They are further directed to pick the number that they think will be closest to 80 percent of the average number chosen by the group. The one who comes closest will receive $100 for his efforts. Stop for a bit and think what number you would pick.

Some in the group might reason that the average number chosen is likely to be 50 and so these people would guess 40, which is 80 percent of this. Others might anticipate that people will guess 40 for this reason and so they would guess 32, which is 80 percent of 40. Still others might anticipate that people will guess 32 for this reason and so they would guess 25.6, which is 80 percent of 32.

If the group continues to play this game, they will gradually learn to engage in ever more iterations of this meta-reasoning about others' reasoning until they all reach the optimal response, which is 0. Since they all want to choose a number equal to 80 percent of the average, the only way they can all do this is by choosing 0, the only number equal to 80 percent of itself. (Choosing 0 leads to what is called the Nash equilibrium of this game. It results when individuals modify their actions until they can no longer benefit from changing them given what the others' actions are.)

The problem of guessing 80 percent of the average guess is a bit like Keynes's description of the investors' task. What makes it tricky is that anyone bright enough to cut to the heart of the problem and guess 0 right away is almost certain to be wrong, since different individuals will engage in different degrees of meta-reasoning about others' reasoning. Some, to increase their chances, will choose numbers a little above or a little below the natural guesses of 40 or 32 or 25.6 or 20.48. There will be some random guesses as well and some guesses of 50 or more. Unless the group is very unusual, few will guess 0 initially.

If a group plays this game only once or twice, guessing the average of all the guesses is as much a matter of reading the others' intelligence and psychology as it is of following an idea to its logical conclusion. By the same token, gauging investors is often as important as gauging investments. And it's likely to be more difficult.

My Pedagogical Cruelty

Other situations, as well, require anticipating others' actions and adapting yours to theirs. Recall, for example, the television show on which contestants had to guess how their spouses would guess they would answer a particular question.

There was also a show on which opposing teams had to guess the most common associations the studio audience had made with a collection of words. Or consider the game in which you have to pick the location in New York City (or simply the local shopping mall) that others would most likely look for you first. You win if the location you pick is chosen by most of the others. Instances of Keynes's beauty contest metaphor are widespread.

As I've related elsewhere, a number of years ago I taught a summer probability course at Temple University. It met every day and the pace was rapid, so to induce my students to keep up with the material I gave a short quiz every day. Applying a perverse idea I'd experimented with in other classes, I placed a little box at the bottom of each exam sheet and a notation next to it stating that students who crossed the box (placed an X in it) would have ten extra points added to their exam scores. A further notation stated that the points would be added only if less than half the class crossed the box. If more than half crossed the box, those crossing it would lose ten points on their exam scores. This practice, I admit, bordered on pedagogical cruelty.

A few brave souls crossed the box on the first quiz and received ten extra points. As the summer wore on, more and more students did so. One day I announced that more than half the students had crossed the box and that those who did had therefore been penalized ten points. Very few students crossed the box on the next exam. Gradually, however, the number crossing it edged up to around 40 percent of the class and stayed there. But it was always a different 40 percent, and it struck me that the calculation a student had to perform to decide whether to cross the box was quite difficult. It was especially so since the class was composed largely of foreign students who, despite my best efforts (which included this little game), seemed to have developed little camaraderie. Without

any collusion that I could discern, the students had to anticipate other students' anticipations of their anticipations in a
convoluted and very skittish self-referential tangle. Dizzying.

I've since learned that W. Brian Arthur, an economist at the
Santa Fe Institute and Stanford University, has long used an essentially identical scenario to describe the predicament of bar
patrons deciding whether or not to go to a popular bar, the experience being pleasant only if the bar is not thronged. An
equilibrium naturally develops whereby the bar rarely becomes
too full. (This almost seems like a belated scientific justification
for Yogi Berra's quip about Toots Shor's restaurant in New
York: "Nobody goes there any more. It's too crowded.")
Arthur proposed the model to clarify the behavior of market
investors who, like my students and the bar patrons, must anticipate others' anticipations of them (and so on). Whether one
buys or sells, crosses the box or doesn't cross, goes to the bar
or doesn't go, depends upon one's beliefs about others' possible
actions and beliefs.

The Consumer Confidence Index, which measures con-
sumers' propensity to consume and their confidence in their
own economic future, is likewise subject to a flighty, reflexive
sort of consensus. Since people's evaluation of their own economic prospects is so dependent on what they perceive others'
prospects to be, the CCI indirectly surveys people's beliefs
about other people's beliefs. ("Consume" and "consumer" are,
in this context, common but unfortunate terms. "Buy," "purchase," "citizen," and "household" are, I think, preferable.)

Common Knowledge,
Jealousy, and Market Sell-Offs

Sizing up other investors is more than a matter of psychology. New logical notions are needed as well. One of them,

"common knowledge," due originally to the economist Robert Aumann, is crucial to understanding the complexity of the stock market and the importance of transparency. A bit of information is common knowledge among a group of people if all parties know it, know that the others know it, know that the others know they know it, and so on. It is much more than "mutual knowledge," which requires only that the parties know the particular bit of information, not that they be aware of the others' knowledge.

As I'll discuss later, this notion of common knowledge is essential to seeing how "subterranean information processing" often underlies sudden bubbles or crashes in the markets, changes that seem to be precipitated by nothing at all and therefore are almost impossible to foresee. It is also relevant to the recent market sell-offs and accounting scandals, but before we get to more realistic accounts of the market, consider the following parable from my book *Once Upon a Number*, which illustrates the power of common knowledge. The story takes place in a benightedly sexist village of uncertain location. In this village there are many married couples and each woman immediately knows when another woman's husband has been unfaithful but not when her own has. The very strict feminist statutes of the village require that if a woman can *prove* her husband has been unfaithful, she must kill him that very day. Assume that the women are statute-abiding, intelligent, aware of the intelligence of the other women, and, mercifully, that they never inform other women of their philandering husbands. As it happens, twenty of the men have been unfaithful, but since no woman can prove her husband has been so, village life proceeds merrily and warily along. Then one morning the tribal matriarch comes to visit from the far side of the forest. Her honesty is acknowledged by all and her word is taken as truth. She warns the assembled villagers that there is at least one philandering husband among them. Once this fact, already known to everyone, becomes *common* knowledge, what happens?

The answer is that the matriarch's warning will be followed by nineteen peaceful days and then, on the twentieth day, by a massive slaughter in which twenty women kill their husbands. To see this, assume there is only one unfaithful husband, Mr. A. Everyone except Mrs. A already knows about him, so when the matriarch makes her announcement, only she learns something new from it. Being intelligent, she realizes that she would know if any other husband were unfaithful. She thus infers that Mr. A is the philanderer and kills him that very day.

Now assume there are two unfaithful men, Mr. A and Mr. B. Every woman except Mrs. A and Mrs. B knows about both these cases of infidelity. Mrs. A knows only of Mr. B's, and Mrs. B knows only of Mr. A's. Mrs. A thus learns nothing from the matriarch's announcement, but when Mrs. B fails to kill Mr. B the first day, she infers that there must be a second philandering husband, who can only be Mr. A. The same holds for Mrs. B who infers from the fact that Mrs. A has not killed her husband on the first day that Mr. B is also guilty. The next day Mrs. A and Mrs. B both kill their husbands.

If there are exactly three guilty husbands, Mr. A, Mr. B, and Mr. C, then the matriarch's announcement would have no visible effect the first day or the second, but by a reasoning process similar to the one above, Mrs. A, Mrs. B, and Mrs. C would each infer from the inaction of the other two of them on the first two days that their husbands were also guilty and kill them on the third day. By a process of mathematical induction we can conclude that if twenty husbands are unfaithful, their intelligent wives would finally be able to prove it on the twentieth day, the day of the righteous bloodbath.

Now if you replace the warning of the matriarch with that provided by, say, an announcement by the Securities and Exchange Commission, the nervousness of the wives with the nervousness of investors, the wives' contentment as long as their own husbands weren't straying with the investors' contentment as long their own companies weren't cooking the books, killing husbands with selling stocks, and the gap between the warning and the killings with the delay between announcement of an investigation and big sell-offs, you can understand how this parable of common knowledge applies to the market.

Note that in order to change the logical status of a bit of information from mutually known to commonly known, there must be an independent arbiter. In the parable it was the matriarch; in the market analogue it was the SEC. If there is no one who is universally respected and believed, the motivating and cleansing effect of warnings is lost.

Happily, unlike the poor husbands, the market is capable of rebirth.

2 | Fear, Greed, and Cognitive Illusions

You don't need to have been a temporarily besotted investor to realize that psychology plays an important and sometimes crucial role in the market, but it helps. By late summer 2000, WCOM had declined to $30 per share, inciting me to buy more. As "inciting" may suggest, my purchases were not completely rational. By this I don't mean that there wasn't a rational basis for investing in WCOM stock. If you didn't look too closely at the problems of overcapacity and the long-distance phone companies' declining revenue streams, you could find reasons to keep buying. It's just that my reasons owed less to an assessment of trends in telecommunications or an analysis of company fundamentals than to an unsuspected gambling instinct and a need to be right. I suffered from "confirmation bias" and searched for the good news, angles, and analyses about the stock while avoiding the less sanguine indications.

Averaging Down or Catching a Falling Knife?

After an increasingly intense, albeit one-sided courtship of the stock (the girl never even sent me a dividend), I married it. As its share price fell, I continued to see only opportunities for gains. Surely, I told myself, the stock had reached its bottom and it was now time to average down by buying the considerably cheaper shares. Of course, for every facile invitation I extended myself to "average down," I ignored an equally facile warning about not attempting to "catch a falling knife." The stale, but prudent adage about not putting too many of one's eggs in the same basket never seemed to push itself very forcefully into my consciousness.

I was also swayed by Salomon Smith Barney's Jack Grubman (possessor, incidentally, of a master's degree in mathematics from Columbia) and other analysts, who ritualistically sprinkled their "strong buys" over the object of my affections. In fact, most brokerage houses in early 2000 rated WCOM a "strong buy," and those that didn't had it as a "buy." It required no great perspicacity to notice that at the time, almost no stock ever received a "sell," much less a "strong sell," and that even "holds" were sparingly bestowed. Maybe, I thought, only environmental companies that manufactured solar-powered flashlights qualified for these latter ratings. Accustomed to grade inflation and to movie, book, and restaurant review inflation, I wasn't taken in by the uniformly positive ratings. Still, just as you can be moved by a television commercial whose saccharine dialogue you are simultaneously ridiculing, part of me gave credence to all those "strong buys."

I kept telling myself that I'd incurred only paper losses and had lost nothing real unless I sold. The stock would come back, and if I didn't sell, I couldn't lose. Did I really believe this? Of course not, but I acted as if I did, and "averaging down" continued to seem like an irresistible opportunity. I believed in the company, but greed and fear were already doing their usual two-step in my head and, in the process, stepping all over my critical faculties.

Emotional Overreactions
and Homo Economicus

Investors can become (to borrow a phrase Alan Greenspan and Robert Shiller made famous) irrationally exuberant, or, changing the arithmetical sign, irrationally despairing. Some of the biggest daily point gains and declines in Nasdaq's history occurred in a single month in early 2000, and the pattern has continued unabated in 2001 and 2002, the biggest point gain since 1987 occurring on July 24, 2002. (The increase in volatility, although substantial, is a little exaggerated since our perception of gains and losses have been distorted by the rise in the indices. A 2 percent drop in the Dow when the market is at 9,000 is 180 points, whereas not too long ago when it was at 3,000, the same percentage drop was only 60 points.) The volatility has come about as the economy has hovered near a recession, as accounting abuses have come to light, as CEO malfeasance has mounted, as the bubble has fizzled, and as people have continued to trade on their own, influenced no doubt by capricious lists of the fifty most beautiful (er . . . , undervalued) stocks.

As with beautiful people and, for that matter, distinguished universities, emotions and psychology are imponderable factors in the market's jumpy variability. Just as beauty and academic quality don't change as rapidly as ad hoc lists and magazine rankings do, so, it seems, the fundamentals of companies don't change as quickly as our mercurial reactions to news about them do.

It may be useful to imagine the market as a fine race car whose exquisitely sensitive steering wheel makes it impossible to drive in a straight line. Tiny bumps in our path cause us to swerve wildly, and we zigzag from fear to greed and back again, from unreasonable gloom to irrational exuberance and back.

Our overreactions are abetted by the all-crisis-all-the-time business media, which brings to mind a different analogy: the reigning theory in cosmology. The inflationary universe hypothesis holds—very, very roughly—that shortly after the Big Bang the primordial universe inflated so fast that all of our visible universe derives from a tiny part of it; we can't see the rest. The metaphor is strained (in fact I just developed carpal tunnel syndrome typing it), but it seems reminiscent of what happens when the business media (as well as the media in general) focus unrelentingly on some titillating but relatively inconsequential bit of news. Coverage of the item expands so fast as to distort the rest of the global village and render it invisible.

Our responses to business news are only one of the ways in which we fail to be completely rational. More generally, we simply don't always behave in ways that maximize our economic well-being. "Homo economicus" is not an ideal toward which many people strive. My late father, for example, was distinctly uneconimicus. I remember him sitting and chuckling on the steps outside our house one autumn night long ago. I asked what was funny and he told me that he had been watching the news and had heard Bob Buhl, a pitcher for the then Milwaukee Braves, answer a TV reporter's question about his off-season plans. "Buhl said he was going to help his father up in Saginaw, Michigan, during the winter." My father laughed again and continued. "And when the reporter asked Buhl what his father did up in Saginaw, Buhl said, 'Nothing at all. He does nothing at all.'"

My father liked this kind of story and his crooked grin lingered on his face. This memory was jogged recently when I was straightening out my office and found a cartoon he had sent me years later. It showed a bum sitting happily on a park bench as a line of serious businessmen traipsed by him. The bum calls out "Who's winning?" Although my father was a salesman, he always seemed less intent on making a sale than on schmoozing with his customers, telling jokes, writing poetry (not all of it doggerel), and taking innumerable coffee breaks.

Everyone can tell such stories, and you would be hard-pressed to find a novel, even one with a business setting, where the characters are all actively pursuing their economic self-interest. Less anecdotal evidence of the explanatory limits of the homo economicus ideal is provided by so-called "ultimatum games." These generally involve two players, one of whom is given a certain amount of money, say $100, by an experimenter, and the other of whom is given a kind of veto. The first player may offer any non-zero fraction of the $100 to the second player, who can either accept or reject it. If he accepts it, he is given whatever amount the first player has offered, and the first player keeps the balance. If he rejects it, the experimenter takes the money back.

Viewing this in rational game-theoretic terms, one would argue that it's in the interest of the second player to accept whatever is offered since any amount, no matter how small, is better than nothing. One would also suspect that the first player, knowing this, would make only tiny offers to the second player. Both suppositions are false. The offers range up to 50 percent of the money involved, and, if deemed too small and therefore humiliating, they are sometimes rejected. Notions of fairness and equality, as well as anger and revenge, seem to play a role.

Behavioral Finance

People's reactions to ultimatum games may be counterproductive, but they are at least clear-eyed. A number of psychologists in recent years have pointed out the countless ways in which we're all subject to other sorts of counterproductive behavior that spring from cognitive blind spots that are analogues, perhaps, of optical illusions. These psychological illusions and foibles often make us act irrationally in a variety of disparate endeavors, not the least of which is investing.

Amos Tversky and Daniel Kahneman are the founders of this relatively new field of study, many of whose early results are reported upon in the classic book *Judgment Under Uncertainty*, edited by them and Paul Slovic. (Kahneman was awarded the 2002 Nobel Prize in economics, and Tversky almost certainly would have shared it had he not died.) Others who have contributed to the field include Thomas Gilovich, Robin Dawes, J. L. Knetschin, and Baruch Fischhoff. Economist Richard Thaler (mentioned in the first chapter) is one of the leaders in applying these emerging insights to economics and finance, and his book *The Winner's Curse*, as well as Gilovich's *How We Know What Isn't So*, are very useful compendiums of recent results.

What makes these results particularly intriguing is the way they illuminate the tactics used, whether consciously or not, by people in everyday life. For example, a favorite ploy of activists of all ideological stripes is to set the terms of a debate by throwing out numbers, which need have little relation to reality to be influential. If you are appalled at some condition, you might want to announce that more than 50,000 deaths each year are attributable to it. By the time people catch up and realize that the number is a couple of orders of magnitude smaller, your cause will be established.

Unfounded financial hype and unrealistic "price targets" have the same effect. Often, it seems, an analyst cites a "price target" for a stock in order to influence investors by putting a number into their heads. (Since the targets are so often indistinguishable from wishes, shouldn't they always be infinite?)

The reason for the success of this hyperbole is that most of us suffer from a common psychological failing. We credit and easily become attached to any number we hear. This tendency is called the "anchoring effect" and it's been demonstrated to hold in a wide variety of situations.

If an experimenter asks people to estimate the population of Ukraine, the size of Avogadro's number, the date of an historical event, the distance to Saturn, or the earnings of XYZ Corporation two years from now, their guesses are likely to be fairly close to whatever figure the experimenter first suggests as a possibility. For example, if he prefaces his request for an estimate of the population of Ukraine with the question—"Is it more or less than 200 million people?"—the subjects' estimates will vary and generally be a bit less than this figure, but still average, say, 175 million people. If he prefaces his request for an estimate with the question—"Is the population of Ukraine more or less than 5 million people?"—the subjects' estimates will vary and this time be a bit more than this figure, but still average, say, 10 million people. The subjects usually move in the right direction from whatever number is presented to them, but nevertheless remain anchored to it.

You might think this is a reasonable strategy for people to follow. They might realize they don't know much about Ukraine, chemistry, history, or astronomy, and they probably believe the experimenter is knowledgeable, so they stick close to the number presented. The astonishing strength of the tendency comes through, however, when the experimenter obtains his preliminary number by some chance means, say by spinning a dial that has numbers around its periphery—300 million, 200 million, 50 million, 5 million, and so on. Say he spins the dial in front of the subjects, points out where it has stopped, and then asks them if the population of Ukraine is more or less than the number at which the dial has stopped.

The subjects' guesses are still anchored to this number even though, one presumes, they don't think the dial knows anything about Ukraine!

Financial numbers are also vulnerable to this sort of manipulation, including price targets and other uncertain future figures like anticipated earnings. The more distant the future the numbers describe, the more it's possible to postulate a huge figure that is justified, say, by a rosy scenario about the exponentially growing need for bandwidth or online airline tickets or pet products. People will discount these estimates, but usually not nearly enough. Some of the excesses of the dot-coms are probably attributable to this effect. On the sell side too, people can paint a dire picture of ballooning debt or shrinking markets or competing technology. Once again, the numbers presented, this time horrific, need not have much to do with reality to have an effect.

Earnings and targets are not the only anchors. People often remember and are anchored to the fifty-two-week high (or low) at which the stock had been selling and continue to base their deliberations on this anchor. I unfortunately did this with WCOM. Having first bought the stock when it was in the forties, I implicitly assumed it would eventually right itself and return there. Later, when I bought more of it in the thirties, twenties, and teens, I made the same assumption.

Another, more extreme form of anchoring (although there are other factors involved) is revealed by investors' focus on whether the earnings that companies announce quarterly meet the estimates analysts have established for them. When companies' earnings fall short by a penny or two per share, investors sometimes react as if this were tantamount to near-bankruptcy. They seem to be not merely anchored to earnings estimates but fetishistically obsessed with them.

Not surprisingly, studies have shown that companies' earnings are much more likely to come in a penny or two above the analysts' average estimate than a penny or two below it. If earnings were figured without regard to analysts' expectations, they'd come in below the average estimate as often as above it. The reason for the asymmetry is probably that companies sometimes "back in" to their earnings. Instead of determining revenues and expenses and subtracting the latter from the former to obtain earnings (or more complicated variants of this), companies begin with the earnings they need and adjust revenues and expenses to achieve them.

Psychological Foibles, A List

The anchoring effect is not the only way in which our faculties are clouded. The "availability error" is the inclination to view any story, whether political, personal, or financial, through the lens of a superficially similar story that is psychologically available. Thus every recent American military involvement is inevitably described somewhere as "another Vietnam." Political scandals are immediately compared to the Lewinsky saga or Watergate, misunderstandings between spouses reactivate old wounds, normal accounting questions bring the Enron-Andersen-WorldCom fiasco to mind, and any new high-tech firm has to contend with memories of the dot-com bubble. As with anchoring, the availability error can be intentionally exploited.

The anchoring effect and availability error are exacerbated by other tendencies. "Confirmation bias" refers to the way we check a hypothesis by observing instances that confirm it and ignoring those that don't. We notice more readily and even diligently search for whatever might confirm our beliefs, and we don't notice as readily and certainly don't look hard for what disconfirms them. Such selective thinking reinforces the anchoring effect: We naturally begin to look for reasons that the arbitrary number presented to us is accurate. If we succumb completely to the confirmation bias, we step over the sometimes fine line separating flawed rationality and hopeless closed-mindedness.

Confirmation bias is not irrelevant to stock-picking. We tend to gravitate toward those people whose take on a stock is similar to our own and to search more vigorously for positive information on the stock. When I visited WorldCom chatrooms, I more often clicked on postings written by people characterizing themselves as "strong buys" than I did on those written by "strong sells." I also paid more attention to WorldCom's relatively small deals with web-hosting companies than to the larger structural problems in the telecommunications industry.

The "status quo bias" (these various biases are generally not independent of each other) also applies to investing. If subjects are told, for example, that they've inherited a good deal of money and then asked which of four investment options (an aggressive stock portfolio, a more balanced collection of equities, a municipal bond fund, or U.S. Treasuries) they would prefer to invest it in, the percentages choosing each are fairly evenly distributed.

Surprisingly, however, if the subjects are told that they've inherited the money but it is already in the form of municipal bonds, almost half choose to keep it in bonds. It's the same with the other three investment options: Almost half elect to keep the money where it is. This inertia is part of the reason so many people sat by while not only their inheritances but their other investments dwindled away. The "endowment effect," another kindred bias, is an inclination to endow one's holdings with more value than they have simply because one holds them. "It's my stock and I love it."

Related studies suggest that passively endured losses induce less regret than losses that follow active involvement. Someone who sticks with an old investment that then declines by 25 percent is less upset than someone who switches into the same investment before it declines by 25 percent. The same fear of regret underlies people's reluctance to trade lottery tickets with friends. They imagine how they'll feel if their original ticket wins.

Minimizing possible regret often plays too large a role in investors' decisionmaking. A variety of studies by Tversky, Kahneman, and others have shown that most people tend to assume less risk to obtain gains than they do to avoid losses. This isn't implausible: Other research suggests that people feel considerably more pain after incurring a financial loss than they do pleasure after achieving an equivalent gain. In the extreme case, desperate fears about losing a lot of money induce people to take enormous risks with their money.

Consider a rather schematic outline of many of the situations studied. Imagine that a benefactor gives $10,000 to everyone in a group and then offers each of them the following choice. He promises to a) give them an additional $5,000 or else b) give them an additional $10,000 or $0, depending on the outcome of a coin flip. Most people choose to receive the additional $5,000. Contrast this with the choice people in a different group make when confronted with a benefactor who gives them each $20,000 and then offers the following choice to each of them. He will a) take from them $5,000 or else b) will take from them $10,000 or $0, depending on the flip of a coin. In this case, in an attempt to avoid any loss, most people choose to flip the coin. The punchline, as it often is, is that the choices offered to the two groups are the same: a sure $15,000 or a coin flip to determine whether they'll receive $10,000 or $20,000.

Alas, I too took more risks to avoid losses than I did to obtain gains. In early October 2000, WCOM had fallen below $20, forcing the CEO, Bernie Ebbers, to sell 3 million shares to pay off some of his investment debts. The WorldCom chatrooms went into one of their typical frenzies and the price dropped further. My reaction, painful to recall, was, "At these prices I can finally get out of the hole." I bought more shares even though I knew better. There was apparently a loose connection between my brain and my fingers, which kept clicking the buy button on my Schwab online account in an effort to avoid the losses that loomed.

Outside of business, loss aversion plays a role as well. It's something of a truism that the attempt to cover up a scandal often leads to a much worse scandal. Although most people know this, attempts to cover up are still common, presumably because, here too, people are much more willing to take risks to avoid losses than they are to obtain gains.

Another chink in our cognitive apparatus is Richard Thaler's notion of "mental accounts," mentioned in the last chapter. "The Legend of the Man in the Green Bathrobe" illustrates this notion compellingly. It is a rather long shaggy dog story, but the gist is that a newlywed on his honeymoon in Las Vegas wakes up in bed and sees a $5 chip left on the dresser. Unable to sleep, he goes down to the casino (in his green bathrobe, of course), bets on a particular number on the roulette wheel, and wins. The 35 to 1 odds result in a payout of $175, which the newlywed promptly bets on the next spin. He wins again and now has more than $6,000. He bets everything on his number a couple more times, continuing until his winnings are in the millions and the casino refuses to accept such a large bet. The man goes to a bigger casino, wins yet again, and now commands hundreds of millions of dollars. He hesitates and then decides to bet it all one more time. This time he loses. In a daze, he stumbles back up to his hotel room where his wife yawns and asks how he did. "Not too bad. I lost $5."

It's not only in casinos and the stock market that we categorize money in odd ways and treat it differently depending on what mental account we place it in. People who lose a $100 ticket on the way to a concert, for example, are less likely to buy a new one than are people who lose $100 in cash on their way to buy the ticket. Even though the amounts are the same in the two scenarios, people in the former one tend to think $200 is too large an expenditure from their entertainment account and so don't buy a new ticket, while people in the latter tend to assign $100 to their entertainment account and $100 to their "unfortunate loss" account and buy the ticket.

In my less critical moments (although not only then) I mentally amalgamate the royalties from this book, whose writing was prompted in part by my investing misadventure, with my WCOM losses. Like corporate accounting, personal accounting can be plastic and convoluted, perhaps even more so since, unlike corporations, we are privately held.

These and other cognitive illusions persist for several reasons. One is that they lead to heuristic rules of thumb that can save time and energy. It's often easier to go on automatic pilot and respond to events in a way that requires little new thinking, not just in scenarios involving eccentric philanthropists and sadistic experimenters. Another reason for the illusions' persistence is that they have, to an extent, become hardwired over the eons. Noticing a rustle in the bush, our primitive ancestors were better off racing away than they were plugging into Bayes' theorem on conditional probability to determine if a threat was really likely.

Sometimes these heuristic rules lead us astray, again not just in business and investing but in everyday life. Early in the fall 2002 Washington, D.C., sniper case, for example, the police arrested a man who owned a white van, a number of rifles, and a manual for snipers. It was thought at the time that there was one sniper and that he owned all these items, so for the purpose of this illustration let's assume that this turned out to be true. Given this and other reasonable assumptions, which is higher—a) the probability that an innocent man would own all these items, or b) the probability that a man who owned all these items would be innocent? You may wish to pause before reading on.

Most people find questions like this difficult, but the second probability would be vastly higher. To see this, let me make up some plausible numbers. There are about 4 million innocent people in the suburban Washington area and, we're assuming, one guilty one. Let's further estimate that ten people (including the guilty one) own all three of the items mentioned above. The first probability—that an innocent man owns all these items—would be 9/4,000,000 or 1 in 400,000. The second probability—that a man owning all three of these items is innocent—would be 9/10. Whatever the actual numbers, these probabilities usually differ substantially. Confusing them is dangerous (to defendants).

Self-Fulfilling Beliefs and Data Mining

Taken to extremes, these cognitive illusions may give rise to closed systems of thought that are immune, at least for a while, to revision and refutation. (Austrian writer and satirist Karl Kraus once remarked, "Psychoanalysis is that mental illness for which it regards itself as therapy.") This is especially true for the market, since investors' beliefs about stocks or a method of picking them can become a self-fulfilling prophecy. The market sometimes acts like a strange beast with a will, if not a mind, of its own. Studying it is not like studying science and mathematics, whose postulates and laws are (in quite different senses) independent of us. If enough people suddenly wake up believing in a stock, it will, for that reason alone, go up in price and justify their beliefs.

A contrived but interesting illustration of a self-fulfilling belief involves a tiny investment club with only two investors and ten possible stocks to choose from each week. Let's assume that each week chance smiles at random on one of the ten stocks the investment club is considering and it rises precipitously, while the week's other nine stocks oscillate within a fairly narrow band.

George, who believes (correctly in this case) that the movements of stock prices are largely random, selects one of the ten stocks by rolling a die (say an icosahedron—a twenty-sided solid—with two sides for each number). Martha, let's assume, fervently believes in some wacky theory, Q analysis. Her choices are therefore dictated by a weekly Q analysis newsletter that selects one stock of the ten as most likely to break out. Although George and Martha are equally likely to pick the lucky stock each week, the newsletter-selected stock will result in big investor gains more frequently than will any other stock.

The reason is simple but easy to miss. Two conditions must be met for a stock to result in big gains for an investor: It must be smiled upon by chance that week and it must be chosen by one of the two investors. Since Martha always picks the newsletter-selected stock, the second condition in her case is always met, so whenever chance happens to favor it, it results in big gains for her. This is not the case with the other stocks. Nine-tenths of the time, chance will smile on one of the stocks that is not newsletter-selected, but chances are George will not have picked that particular one, and so it will seldom result in big gains for him. One must be careful in interpreting this, however. George and Martha have equal chances of pulling down big gains (10 percent), and each stock of the ten has an equal chance of being smiled upon by chance (10 percent), but the newsletter-selected stock will achieve big gains much more often than the randomly selected ones.

Reiterated more numerically, the claim is that 10 percent of the time the newsletter-selected stock will achieve big gains for Martha, whereas each of the ten stocks has only a 1 percent chance of both achieving big gains *and* being chosen by George. Note again that two things must occur for the newsletter-selected stock to achieve big gains: Martha must choose it, which happens with probability 1, and it must be the stock that chance selects, which happens with probability 1/10th. Since one multiplies probabilities to determine the likelihood that several independent events occur, the probability of both these events occurring is $1 \times 1/10$, or 10 percent. Likewise, two things must occur for any particular stock to achieve big gains via George: George must choose it, which occurs with probability 1/10th, and it must be the stock that chance selects, which happens with probability 1/10th. The product of these two probabilities is 1/100th or 1 percent.

Nothing in this thought experiment depends on there being only two investors. If there were one hundred investors, fifty of whom slavishly followed the advice of the newsletter and fifty of whom chose stocks at random, then the newsletter-selected stocks would achieve big gains for their investors eleven times as frequently as any particular stock did for its investors. When the newsletter-selected stock is chosen by chance and happens to achieve big gains, there are fifty-five winners, the fifty believers in the newsletter and five who picked the same stock at random. When any of the other nine stocks happens to achieve big gains, there are, on average, only five winners.

In this way a trading strategy, if looked at in a small population of investors and stocks, can give the strong illusion that it is effective when only chance is at work.

"Data mining," the scouring of databases of investments, stock prices, and economic data for evidence of the effectiveness of this or that strategy, is another example of how an inquiry of limited scope can generate deceptive results. The problem is that if you look hard enough, you will always find some seemingly effective rule that resulted in large gains over a certain time span or within a certain sector. (In fact, inspired by the British economist Frank Ramsey, mathematicians over the last half century have proved a variety of theorems on the inevitability of some kind of order in large sets.) The promulgators of such rules are not unlike the believers in bible codes. There, too, people searched for coded messages that seemed to be meaningful, not realizing that it's nearly impossible for there not to be some such "messages." (This is trivially so if you search in a book that has a chapter 11, conveniently foretelling many companies' bankruptcies.)

People commonly pore over price and trade data attempting to discover investment schemes that have worked in the past. In a reductio ad absurdum of such unfocused fishing for associations, David Leinweber in the mid–90s exhaustively searched the economic data on a United Nations CD-ROM and found that the best predictor of the value of the S&P 500 stock index was—a drum roll here—butter production in Bangladesh. Needless to say, butter production in Bangladesh has probably not remained the best predictor of the S&P 500. Whatever rules and regularities are discovered within a sample must be applied to new data if they're to be accorded any limited credibility. You can always arbitrarily define a class of stocks that in retrospect does extraordinarily well, but will it continue to do so?

I'm reminded of a well-known paradox devised (for a different purpose) by the philosopher Nelson Goodman. He selected an arbitrary future date, say January 1, 2020, and defined an object to be "grue" if it is green and the time is before January 1, 2020, or if it is blue and the time is after January 1, 2020. Something is "bleen," on the other hand, if it is blue and the time is before that date or if it is green and the time is after that date. Now consider the color of emeralds. All emeralds examined up to now (2002) have been green. We therefore feel confident that all emeralds are green. But all emeralds so far examined are also grue. It seems that we should be just as confident that all emeralds are grue (and hence blue beginning in 2020). Are we?

A natural objection is that these color words grue and bleen are very odd, being defined in terms of the year 2020. But were there aliens who speak the grue-bleen language, they could make the same charge against us. "Green," they might argue, is an arbitrary color word, being defined as grue before 2020 and bleen afterward. "Blue" is just as odd, being bleen before 2020 and grue from then on. Philosophers have not convincingly shown what exactly is wrong with the terms grue and bleen, but they demonstrate that even the abrupt failure of a regularity to hold can be accommodated by the introduction of new weasel words and ad hoc qualifications.

In their headlong efforts to discover associations, data miners are sometimes fooled by "survivorship bias." In market usage this is the tendency for mutual funds that go out of business to be dropped from the average of all mutual funds. The average return of the surviving funds is higher than it would be if all funds were included. Some badly performing funds become defunct, while others are merged with better-performing cousins. In either case, this practice skews past returns upward and induces greater investor optimism about future returns. (Survivorship bias also applies to stocks, which come and go over time, only the surviving ones making the statistics on performance. WCOM, for example, was unceremoniously replaced on the S&P 500 after its steep decline in early 2002.)

The situation is rather like that of schools that allow students to drop courses they're failing. The grade point averages of schools with such a policy are, on average, higher than those of schools that do not allow such withdrawals. But these inflated GPAs are no longer a reliable guide to students' performance.

Finally, taking the meaning of the term literally, survivorship bias makes us all a bit more optimistic about facing crises. We tend to see only those people who survived similar crises. Those who haven't are gone and therefore much less visible.

Rumors and Online Chatrooms

Online chatrooms are natural laboratories for the observation of illusions and distortions, although their psychology is more often brutally basic than subtly specious. While spellbound by WorldCom, I would spend many demoralizing, annoying, and engaging hours compulsively scouring the various World-Com discussions at Yahoo! and RagingBull. Only a brief visit to these sites is needed to see that a more accurate description of them would be rantrooms.

Once someone dons a screen name, he (the masculine pronoun, I suspect, is almost always appropriate) usually dispenses with grammar, spelling, and most conventional standards of polite discourse. Other people become morons, idiots, and worse. A poster's references to the stock, if he's shorting it (selling shares he doesn't have in the hope that he can buy them back when the price goes down), put a burden on one's ability to decode scatological allusions and acronyms. Any expression of pain at one's losses is met with unrelenting scorn and sarcasm; ostensibly genuine musings about suicide are no exception. A suicide threat in April 2002, lamenting the loss of house, family, and job because of WCOM, drew this response: "You sad sack loser. Die. You might want to write a note too in case the authorities and your wife don't read the Yahoo! chatrooms."

People who characterize themselves as sellers are generally (but not always) more vituperative than those claiming to be buyers. Some of the regulars appear genuinely interested in discussing the stock rationally, imparting information, and exchanging speculation. A few seem to know a lot, many are devotees of various outlandish conspiracy theories, including the usual anti-Semitic sewage, and even more are just plain clueless, asking, for example, why they "always put that slash between the P and the E in P/E, and is P price or profit." There were also many discussions that had nothing directly to do with the stock. One that I remember fondly was about someone who called a computer help desk because his computer didn't work. It turned out that he had plugged the computer and all his peripheral devices into his surge protector, which he then plugged into itself. The connection with whatever company was being discussed I've forgotten.

Taking advice from such an absurdly skewed sample of posters is silly, of course, but the real-time appeal of the sites is akin to overhearing gossip about a person you're interested in. It's likely to be false, spun, or overstated, but it still holds a certain fascination. Another analogy is to listening to police radio and getting a feel for the raw life and death on the streets.

Chatroom denizens form little groups that spend a lot of time excoriating, but not otherwise responding to, opposing groups. They endorse each other's truisms and denounce those of the others. When WorldCom purchased a small company or had a reversal in its Brazilian operations, this was considered big news. It was not nearly as significant, however, as an analyst changing his recommendation from a strong buy to a buy or vice versa. If you filter out the postings drenched in anger and billingsgate, you find most of the biases mentioned above demonstrated on a regular basis. The posters are averse to risk, anchored to some artificial number, addicted to circular thought, impressed by data mining, or all of the above.

Most boards I visited had a higher percentage of rational posters than did WorldCom's. I remember visiting the Enron board and reading rumors of the bogus deals and misleading accounting practices that eventually came to light. Unfortunately, since there are always rumors of every conceivable and contradictory sort (sometimes posted by the same individual), one cannot conclude anything from their existence except that they're likely to contribute to feelings of hope, fear, anger, and anxiety.

Pump and Dump, Short and Distort

The rumors are often associated with market scams that exploit people's normal psychological reactions. Many of these reactions are chronicled in Edwin Lefevre's 1923 classic novel, *Reminiscences of a Stock Operator,* but the standard "pump and dump" is an illegal practice that has gained new life on the Internet. Small groups of individuals buy a stock and tout it in a misleading hyperbolic way (that is, pump it). Then when its price rises in response to this concerted campaign, they sell it at a profit (dump it). The practice works best in bull markets when people are most susceptible to greed. It is also most effective when used on thinly traded stocks where a few buyers can have a pronounced effect.

Even a single individual with a fast Internet connection and a lot of different screen names can mount a pump and dump operation. Just buy a small stock from an online broker, then visit the chatroom where it's discussed. Post some artful innuendoes or make some outright phony claims and then back yourself up with one of your pseudonyms. You can even maintain a "conversation" among your various screen names, each salivating over the prospects of the stock. Then just wait for it to move up and sell it quickly when it does.

A fifteen-year-old high school student in New Jersey was arrested for successfully pumping and dumping after school. It's hard to gauge how widespread the practice is since the perpetrators generally make themselves invisible. I don't think it's rare, especially since there are gradations in the practice, ranging from organized crime telephone banks to conventional brokers inveigling gullible investors.

In fact, the latter probably constitute a vastly bigger threat. Being a stock analyst used to be a thoroughly respectable profession, and for most practitioners no doubt it still is. Unfortunately, however, there seem to be more than a few whose fervent desire to obtain the investment banking fees associated with underwriting, mergers, and the other quite lucrative practices induces them to shade their analyses—and "shade" may be a kind verb—so as not to offend the companies they're both analyzing and courting. In early 2002, there were well-publicized stories of analysts at Merrill Lynch exchanging private emails deriding a stock that they were publicly touting. Six other brokerage houses were accused of similar wrongdoing.

Even more telling were records from Salomon Smith Barney subpoenaed by Congress indicating that executives at companies generating large investment fees often *personally* received huge dollops of companies' initial public offerings. Not open to ordinary investors, these hot, well-promoted offerings quickly rose in value and their quick sale generated immediate profits. Bernie Ebbers was reported to have received, between 1996 and 2000, almost a million shares of IPOs worth more than $11 million. The $1.4 billion settlement between several big brokerage houses and the government announced in December 2002 left little doubt that the practice was not confined to Ebbers and Salomon.

In retrospect it now seems that some analysts' ratings weren't much more credible than the ubiquitous email invitations from people purporting to be Nigerian government officials in need of a little seed money. The usual claim is that the money will enable them and their gullible respondents to share in enormous, but frozen foreign accounts.

The bear market analogue to pumping and dumping is shorting and distorting. Instead of buying, touting, and selling on the jump in price, shorters and distorters sell, lambast, and buy on the decline in price.

They first short-sell the stock in question. As mentioned, that is the practice of selling shares one doesn't own in the hope that the price of the shares will decline when it comes time to pay the broker for the borrowed shares. (Short-selling is perfectly legal and also serves a useful purpose in maintaining markets and limiting risk.) After short-selling the stock, the scamsters lambast it in a misleading hyperbolical way (that is, distort its prospects). They spread false rumors of writedowns, unsecured debts, technology problems, employee morale, legal proceedings. When the stock's price declines in response to this concerted campaign, they buy the shares at the lower price and keep the difference.

Like its bull-market counterpart, shorting and distorting works best on thinly traded stocks. It is most effective in a bear market when people are most susceptible to fear and anxiety. Online practitioners, like pumpers and dumpers, use a variety of screen names, this time to create the illusion that something catastrophic is about to befall the company in question. They also tend to be nastier toward investors who disagree with them than are pumpers and dumpers, who must maintain a sunny, confident air. Again there are gradations in the practice and it sometimes seems indistinguishable from some fairly conventional practices of brokerage houses and hedge funds.

Even large stocks like WCOM (with 3 billion outstanding shares) can be affected by such shorters and distorters although they must be better placed than the dermatologically challenged isolates who usually carry on the practice. I don't doubt that there was much shorting of WCOM during its long descent, although given what's come to light about the company's accounting, "short and report" is a more faithful description of what occurred.

Unfortunately, after Enron, WorldCom, Tyco, and the others, even an easily generated whiff of malfeasance can cause investors to sell first and ask questions later. As a result, many worthy companies are unfairly tarred and their investors unnecessarily burned.

3 | Trends, Crowds, and Waves

As a predictor of stock prices, psychology goes only so far. Many investors subscribe to "technical analysis," an approach generally intent on discerning the short-term direction of the market via charts and patterns and then devising rules for pursuing it. Adherents of technical analysis, which is not all that technical and would more accurately be termed "trend analysis," believe that "the trend is their friend," that "momentum investing" makes sense, that crowds should be followed. Whatever the validity of these beliefs and of technical analysis in general (and I'll get to this shortly), I must admit to an a priori distaste for the herdish behavior it often seems to counsel: Figure out where the pack is going and follow it. It was this distaste, perhaps, that prevented me from selling WCOM and that caused me to sputter continually to myself that the company was the victim of bad public relations, investor misunderstanding, media bashing, anger at the CEO, a poisonous business climate, unfortunate timing, or panic selling. In short, I thought the crowd was wrong and hated the idea that it must be obeyed. As I slowly learned, however, disdaining the crowd is sometimes simply hubris.

Technical Analysis:
Following the Followers

My own prejudices aside, the justification for technical analysis is murky at best. To the extent there is one, it most likely derives from psychology, perhaps in part from the Keynesian idea of conventionally anticipating the conventional response, or perhaps from some as yet unarticulated systemic interactions. "Unarticulated" is the key word here: The quasi-mathematical jargon of technical analysis seldom hangs together as a coherent theory. I'll begin my discussion of it with one of its less plausible manifestations, the so-called Elliott wave theory.

Ralph Nelson Elliott famously believed that the market moved in waves that enabled investors to predict the behavior of stocks. Outlining his theory in 1939, Elliott wrote that stock prices move in cycles based upon the Fibonacci numbers (1, 2, 3, 5, 8, 13, 21, 34, 55, 89, . . . , each successive number in the sequence being the sum of the two previous ones). Most commonly the market rises in five distinct waves and declines in three distinct waves for obscure psychological or systemic reasons. Elliott believed as well that these patterns exist at many levels and that any given wave or cycle is part of a larger one and contains within it smaller waves and cycles. (To give Elliott his due, this idea of small waves within larger ones having the same structure does seem to presage mathematician Benoit Mandelbrot's more sophisticated notion of a fractal, to which I'll return later.) Using Fibonacci-inspired rules, the investor buys on rising waves and sells on falling ones.

The problem arises when these investors try to identify where on a wave they find themselves. They must also decide whether the larger or smaller cycle of which the wave is inevitably a part may temporarily be overriding the signal to buy or sell. To save the day, complications are introduced into the theory, so many, in fact, that the theory soon becomes incapable of being falsified. Such complications and unfalsifiability are reminiscent of the theory of biorhythmns and many other pseudosciences. (Biorhythm theory is the idea that various aspects of one's life follow rigid periodic cycles that begin at birth and are often connected to the numbers 23 and 28, the periods of some alleged male and female principles, respectively.) It also brings to mind the ancient Ptolemaic system of describing the planets' movements, in which more and more corrections and ad hoc exceptions had to be created to make the system jibe with observation. Like most other such schemes, Elliott wave theory founders on the simple question: Why should anyone expect it to work?

For some, of course, what the theory has going for it is the mathematical mysticism associated with the Fibonacci numbers, any two adjacent ones of which are alleged to stand in an aesthetically appealing relation. Natural examples of Fibonacci series include whorls on pine cones and pineapples; the number of leaves, petals, and stems on plants; the numbers of left and right spirals in a sunflower; the number of rabbits in succeeding generations; and, insist Elliott enthusiasts, the waves and cycles in stock prices.

It's always pleasant to align the nitty-gritty activities of the market with the ethereal purity of mathematics.

The Euro and the Golden Ratio

Before moving on to less barren financial theories, I invite you to consider a brand new instance of financial numerology. An email from a British correspondent apprised me of an interesting connection between the euro-pound and pound-euro exchange rates on March 19, 2002.

To appreciate it, one needs to know the definition of the golden ratio from classical Greek mathematics. (Those for whom the confluence of Greek, mathematics, and finance is a bit much may want to skip to the next section.) If a point on a straight line divides the line so that the ratio of the longer part to the shorter is equal to the ratio of the whole to the longer part, the point is said to divide the line in a golden ratio. Rectangles whose length and width stand in a golden ratio are also said to be golden, and many claim that rectangles of this shape, for example, the facade of the Parthenon, are particularly pleasing to the eye. Note that a 3-by-5 card is almost a golden rectangle since 5/3 (or 1.666 . . .) is approximately equal to (5 + 3)/5 (or 1.6).

The value of the golden ratio, symbolized by the Greek letter phi, is 1.618 . . . (the number is irrational and so its decimal representation never repeats). It is not difficult to prove that phi has the striking property that it is exactly equal to 1 plus its reciprocal (the reciprocal of a number is simply 1 divided by the number). Thus 1.618 . . . is equal to 1 + 1/1.618 . . . .

This odd fact returns us to the euro and the pound. An announcer on the BBC on the day in question, March 19, 2002, observed that the exchange rate for 1 pound sterling was 1 euro and 61.8 cents (1.618 euros) and that, lo and behold, this meant that the reciprocal exchange rate for 1 euro was 61.8 pence (.618 pounds). This constituted, the announcer went on, "a kind of symmetry." The announcer probably didn't realize how profound this symmetry was.

In addition to the aptness of "golden" in this financial context, there is the following well-known relation between the golden ratio and the Fibonacci numbers. The ratio of any Fibonacci number to its predecessor is close to the golden ratio of 1.618 . . . , and the bigger the numbers involved, the closer the two ratios become. Consider again, the Fibonacci numbers, 1, 2, 3, 5, 8, 13, 21, 34, 55, ... The ratios, 5/3, 8/5,

13/8, 21/13, ... , of successive Fibonacci numbers approach the golden ratio of 1.618 ... !

There's no telling how an Elliott wave theorist dabbling in currencies at the time of the above exchange rate coincidence would have reacted to this beautiful harmony between money and mathematics. An unscrupulous, but numerate hoaxer might have even cooked up some flapdoodle sufficiently plausible to make money from such a "cosmic" connection.

The story could conceivably form the basis of a movie like *Pi*, since there are countless odd facts about phi that could be used to give various investing schemes a superficial plausibility. (The protagonist of *Pi* was a numerologically obsessed mathematician who thought he'd found the secret to just about everything in the decimal expansion of pi. He was pursued by religious zealots, greedy financiers, and others. The only sane character, his mentor, had a stroke, and the syncopated black-and-white cinematography was anxiety-inducing. Appealing as it was, the movie was mathematically nonsensical.) Unfortunately for investors and mathematicians alike, the lesson again is that more than beautiful harmonies are needed to make money on Wall Street. And *Phi* can't match the cachet of *Pi* as a movie title either.

Moving Averages, Big Picture

People, myself included, sometimes ridicule technical analysis and the charts associated with it in one breath and then in the next reveal how much in (perhaps unconscious) thrall to these ideas they really are. They bring to mind the old joke about the man who complains to his doctor that his wife has for several years believed she's a chicken. He would have sought help sooner, he says, "but we needed the eggs." Without reading

too much into this story except that we do sometimes seem to
need the notions of technical analysis, let me finally proceed
to examine some of these notions.

Investors naturally want to get a broad picture of the
movement of the market and of particular stocks, and for this
the simple technical notion of a moving average is helpful.
When a quantity varies over time (such as the stock price of a
company, the noontime temperature in Milwaukee, or the
cost of cabbage in Kiev), one can, each day, average its values
over, say, the previous 200 days. The averages in this sequence vary and hence the sequence is called a moving average, but the value of such a moving average is that it doesn't
move nearly as much as the stock price itself; it might be
termed the phlegmatic average.

For illustration, consider the three-day moving average of a
company whose stock is very volatile, its closing prices on
successive days being: 8, 9, 10, 5, 6, 9. On the day the stock
closed at 10, its three-day moving average was $(8 + 9 + 10)/3$
or 9. On the next day, when the stock closed at 5, its three-
day moving average was $(9 + 10 + 5)/3$ or 8. When the stock
closed at 6, its three-day moving average was $(10 + 5 + 6)/3$
or 7. And the next day, when it closed at 9, its three-day moving average was $(5 + 6 + 9)/3$ or 6.67.

If the stock oscillates in a very regular way and you are
careful about the length of time you pick, the moving average
may barely move at all. Consider an extreme case, the twenty-
day moving average of a company whose closing stock prices
oscillate with metronomic regularity. On successive days they
are: 51, 52, 53, 54, 55, 54, 53, 52, 51, 50, 49, 48, 47, 46, 45,
46, 47, 48, 49, **50**, 51, 52, 53, and so on, moving up and
down around a price of 50. The twenty-day moving average
on the day marked in bold is 50 (obtained by averaging the
20 numbers up to and including it). Likewise, the twenty-day
moving average on the next day, when the stock is at 51, is

also 50. It's the same for the next day. In fact, if the stock price oscillates in this regular way and repeats itself every twenty days, the twenty-day moving average is always 50.

There are variations in the definition of moving averages (some weight recent days more heavily, others take account of the varying volatility of the stock), but they are all designed to smooth out the day-to-day fluctuations in a stock's price in order to give the investor a look at broader trends. Software and online sites allow easy comparison of the stock's daily movements with the slower-moving averages.

Technical analysts use the moving average to generate buy-sell rules. The most common such rule directs you to buy a stock when it exceeds its X-day moving average. Context determines the value of X, which is usually 10, 50, or 200 days. Conversely, the rule directs you to sell when the stock falls below its X-day moving average. With the regularly oscillating stock above, the rule would not lead to any gains or losses. It would call for you to buy the stock when it moves from 50, its moving average, to 51, and for you to sell it when it moves from 50 to 49. In the previous example of the three-day moving average, the rule would require that you buy the stock at the end of the third day and sell it at the end of the fourth, leading in this particular case to a loss.

The rule can work well when a stock fluctuates about a long-term upward- or downward-sloping course. The rationale for it is that trends should be followed, and that when a stock moves above its X-day moving average, this movement signals that a bullish trend has begun. Conversely, when a stock moves below its X-day moving average, the movement signals a bearish trend. I reiterate that mere upward (downward) movement of the stock is not enough to signal a buy (sell) order; a stock must move above (below) its moving average.

Alas, had I followed any sort of moving average rule, I would have been out of WCOM, which moved more or less steadily downhill for almost three years, long before I lost most of my investment in it. In fact, I never would have bought it in the first place. The security guard mentioned in chapter 1 did, in effect, use such a rule to justify the sale of the stocks in his pension plan.

There are a few studies, which I'll get to later, suggesting that a moving average rule is sometimes moderately effective. Even so, however, there are several problems. One is that it can cost you a lot in commissions if the stock price hovers around the moving average and moves through it many times in both directions. Thus you have to modify the rule so that the price must move above or below its moving average by a non-trivial amount. You must also decide whether to buy at the end of the day the price exceeds the moving average or at the beginning of the next day or later still.

You can mine the voluminous time-series data on stock prices to find the X that has given the best returns for adhering to the X-day moving average buy-sell rule. Or you can complicate the rule by comparing moving averages over different intervals and buying or selling when these averages cross each other. You can even adapt the idea to day trading by using X-minute moving averages defined in terms of the mathematical notion of an integral. Optimal strategies can always be found after the fact. The trick is getting something that will work in the future; everyone's very good at predicting the past. This brings us to the most trenchant criticism of the moving-average strategy. If the stock market is efficient, that is, if information about a stock is almost instantaneously incorporated into its price, then any stock's future moves will be determined by random external events. Its past behavior, in particular its moving average, is irrelevant, and its future movement is unpredictable.

Of course, the market may not be all that efficient. There'll be much more on this question in later chapters.

Resistance and Support and All That

Two other important ideas from technical analysis are resistance and support levels. The argument for them assumes that people usually remember when they've been burned, insulted, or left out; in particular, they remember what they paid, or wish they had paid, for a stock. Assume a stock has been selling for $40 for a while and then drops to $32 before slowly rising again. The large number of people who bought it around $40 are upset and anxious to recoup their losses, so if the stock moves back up to $40, they're likely to sell it, thereby driving the price down again. The $40 price is termed a resistance level and is considered an obstacle to further upward movement of the stock price.

Likewise, investors who considered buying at $32 but did not are envious of those who did buy at that price and reaped the 25 percent returns. They are eager to get these gains, so if the stock falls back to $32, they're likely to buy it, driving the price up again. The $32 price is termed a support level and is considered an obstacle to further downward movement.

Since stocks often seem to meander between their support and resistance levels, one rule followed by technical analysts is to buy the stock when it "bounces" off its support level and sell it when it "bumps" up against its resistance level. The rule can, of course, be applied to the market as a whole, inducing investors to wait for the Dow or the S&P to definitively turn up (or down) before buying (or selling).

Since chartists tend to view support levels as shaky, often temporary, floors and resistance levels as slightly stronger, but still temporary, ceilings, there is a more compelling rule involving these notions. It instructs you to buy the stock if the rising price breaks through the resistance level and to sell it if the falling price breaks through the support level. In both these cases breaking through indicates that the stock has moved out of its customary channel and the rule counsels investors to follow the new trend.

As with the moving-average rules, there are a few studies that indicate that resistance-support rules sometimes lead to moderate increases in returns. Against this there remains the perhaps dispiriting efficient-market hypothesis, which maintains that past prices, trends, and resistance and support levels provide no evidence about future movements.

Innumerable variants of these rules exist and they can be combined in ever more complicated ways. The resistance and support levels can change and trend up or down in a channel or with the moving average, for example, rather than remain fixed. The rules can also be made to take account of variations in a stock's volatility as well.

These variants depend on price patterns that often come equipped with amusing names. The "head and shoulders" pattern, for example, develops after an extended upward trend. It is comprised of three peaks, the middle and highest one being the head, and the smaller left and right ones (earlier and later ones, that is) being the shoulders. After falling below the right shoulder and breaking through the support line connecting the lows on either side of the head, the stock price has, technical chartists aver, reversed direction and a downward trend has begun, so sell.

Similar metaphors describe the double-bottom trend reversal. It develops after an extended downward trend and is comprised of two successive troughs or bottoms with a small peak between them. After bouncing off the second bottom, the stock has, technical chartists again aver, reversed direction and an upward trend has begun, so buy.

These are nice stories, and technical analysts tell them with great earnestness and conviction. Even if everyone told the same stories (and they don't), why should they be true? Presumably the rationale is ultimately psychological or perhaps sociological or systemic, but exactly what principles justify these beliefs? Why not triple or quadruple bottoms? Or two heads and shoulders? Or any of innumerable other equally plausible, equally risible patterns? What combination of psychological, financial, or other principles has sufficient specificity to generate effective investment rules?

As with Elliott waves, scale is an issue. If we go to the level of ticks, we can find small double bottoms and little heads and tiny shoulders all over. We find them also in the movement of broad market indices. And do these patterns mean for the market as whole what they are purported to mean for individual stocks? Is the "double-dip" recession discussed in early 2002 simply a double bottom?

Predictability and Trends

I often hear people swear that they make money using the rules of technical analysis. Do they really? The answer, of course, is that they do. People make money using all sorts of strategies, including some involving tea leaves and sunspots. The real question is: Do they make *more* money than they would investing in a blind index fund that mimics the performance of the market as a whole? Do they achieve excess returns? Most financial theorists doubt this, but there is some tantalizing evidence for the effectiveness of momentum strategies or short-term trend-following. Economists Narasimhan Jegadeesh and Sheridan Titman, for example, have written several papers arguing that momentum strategies result in moderate excess returns and that, having done so over the years, their success is not the result of data mining. Whether this alleged profitability—many dispute it—is due to overreactions among investors or to the short-term persistence of the impact of companies' earnings reports, they don't say. They do seem to point to behavioral models and psychological factors as relevant.

William Brock, Josef Lakonishok, and Blake LeBaron have also found some evidence that rules based on moving averages and the notions of resistance and support are moderately effective. They focus on the simplest rules, but many argue that their results have not been replicated on new stock data.

More support for the existence of technical exploitability comes from Andrew Lo, who teaches at M.I.T., and Craig MacKinlay, from the Wharton School. They argue in their book, *A Non-Random Walk Down Wall Street*, that in the short run overall market returns are, indeed, slightly positively correlated, much like the local weather. A hot, sunny day is a bit more likely to be followed by another one, just as a good week in the market is a bit more likely to be followed by another one. Likewise for rainy days and bad markets. Employing state-of-the-art tools, Lo and MacKinlay also claim that in the long term the prognosis changes: Individual stock prices display a slight negative correlation. Winners are a bit more likely to be losers three to five years hence and vice versa.

They also bring up an interesting theoretical possibility. Weeding out some of the details, let's assume for the sake of the argument (although Lo and MacKinlay don't) that the thesis of Burton Malkiel's classic book, *A Random Walk Down Wall Street*, is true and that the movement of the market as a whole is entirely random. Let's also assume that each stock, when its fluctuations are examined in isolation, moves randomly. Given these assumptions it would nevertheless still be possible that the price movements of, say, 5 percent of stocks accurately predict the price movements of a different 5 percent of stocks one week later.

The predictability comes from cross-correlations over time between stocks. (These associations needn't be causal, but might merely be brute facts.) More concretely, let's say stock

X, when looked at in isolation, fluctuates randomly from week to week, as does stock Y. Yet if X's price this week often predicts Y's next week, this would be an exploitable opportunity and the strict random-walk hypothesis would be wrong. Unless we delved deeply into such possible cross-correlations among stocks, all we would see would be a randomly fluctuating market populated by randomly fluctuating stocks. Of course, I've employed the typical mathematical gambit of considering an extreme case, but the example does suggest that there may be relatively simple elements of order in a market that appears to fluctuate randomly.

There are other sorts of stock price anomalies that can lead to exploitable opportunities. Among the most well-known are so-called calendar effects whereby the prices of stocks, primarily small-firm stocks, rise disproportionately in January, especially during the first week of January. (The price of WCOM rose significantly in January 2001, and I was hoping this rise would repeat itself in January 2002. It didn't.) There has been some effort to explain this by citing tax law concerns that end with the close of the year, but the effect also seems to hold in countries with different tax laws. Moreover, unusual returns (good or bad) occur not only at the turn of the year, but, as Richard Thaler and others have observed, at the turn of the month, week, and day as well as before holidays. Again, poorly understood behavioral factors seem to be involved.

Technical Strategies and Blackjack

Most academic financial experts believe in some form of the random-walk theory and consider technical analysis almost indistinguishable from a pseudoscience whose predictions are either worthless or, at best, so barely discernibly better than chance as to be unexploitable because of transaction costs.

I've always leaned toward this view, but I'll reserve my more nuanced judgment for later in the book. In the meantime, I'd like to point out a parallel between market strategies such as technical analysis in one of its many forms and blackjack strategies. (There are, of course, great differences too.)

Blackjack is the only casino game of chance whose outcomes depend on past outcomes. In roulette, the previous spins of the wheel have no effect on future spins. The probability of red on the next spin is 18/38, even if red has come up on the five previous spins. The same is true with dice, which are totally lacking in memory. The probability of rolling a 7 with a pair of dice is 1/6, even if the four previous rolls have not resulted in a single 7. The probability of six reds in a row is $(18/38)^6$; the probability of five 7s in a row is $(1/6)^5$. Each spin and each roll are independent of the past.

A game of blackjack, on the other hand, is sensitive to its past. The probability of drawing two aces in a row from a deck of cards is not $(4/52 \times 4/52)$ but rather $(4/52 \times 3/51)$. The second factor, 3/51, is the probability of choosing another ace given that the first card chosen was an ace. In the same way the probability that a card drawn from a deck will be a face card (jack, queen, or king) given that only three of the thirty cards drawn so far have been face cards is not 12/52, but a much higher 9/22.

This fact—that (conditional) probabilities change according to the composition of the remaining portion of the deck—is the basis for various counting strategies in blackjack that involve keeping track of how many cards of each type have already been drawn and increasing one's bet size when the odds are (occasionally and slightly) in one's favor. Some of these strategies, followed carefully, do work. This is evidenced by the fact that some casinos supply burly guards free of charge to abruptly escort successful counting practitioners from the premises.

The vast majority of people who try these strategies (or, worse, others of their own devising) lose money. It would make no sense, however, to point to the unrelenting average losses of blackjack players and maintain that this proves that there is no effective betting strategy for playing the game.

Blackjack is much simpler than the stock market, of course, which depends on vastly more factors as well as on the actions and beliefs of other investors. But the absence of conclusive evidence for the effectiveness of various investing rules, technical or otherwise, does not imply that no effective rules exist. If the market's movements are not completely random, then it has a kind of memory within it, and investing rules depending on this memory might be effective. Whether they would remain so if widely known is very dubious, but that is another matter.

Interestingly, if there were an effective technical trading strategy, it wouldn't need any convincing rationale. Most investors would be quite pleased to use it, as most blackjack players use the standard counting strategy, without understanding why it works. With blackjack, however, there is a compelling mathematical explanation for those who care to study it. By contrast an effective technical trading strategy might be found that was beyond the comprehension not only of the people using it but of everyone. It might simply work, at least temporarily. In Plato's allegory of the cave the benighted see only the shadows on the wall of the cave and not the real objects behind them that are causing the shadows. If they were really predictive, investors would be quite content with the shadows alone and would simply take the cave to be a bargain basement.

The next segment is a bit of a lark. It offers a suggestive hint for developing a novel and counterintuitive investment strategy that has a bit of the feel of technical analysis.

Winning Through Losing?

The old joke about the store owner losing money on every sale but making it up in volume may have a kernel of truth to it. An interesting new paradox by Juan Parrondo, a Spanish physicist, brings the joke to mind. It deals with two games, each of which results in steady losses over time. When these games are played in succession in random order, however, the result is a steady gain. Bad bets strung together to produce big winnings—very strange indeed!

To understand Parrondo's paradox, let's switch from a financial to a spatial metaphor. Imagine you are standing on stair 0, in the middle of a very long staircase with 1,001 stairs numbered from –500 to 500 (–500, –499, –498, . . . , –4, –3, –2, –1, 0, 1, 2, 3, 4, . . . , 498, 499, 500). You want to go up rather than down the staircase and which direction you move depends on the outcome of coin flips. The first game—let's call it game S—is very Simple. You flip a coin and move up a stair whenever it comes up heads and down a stair whenever it comes up tails. The coin is slightly biased, however, and comes up heads 49.5 percent of the time and tails 50.5 percent. It's clear that this is not only a boring game but a losing one. If you played it long enough, you would move up and down for a while, but almost certainly you would eventually reach the bottom of the staircase.

The second game—let's continue to wax poetic and call it game C—is more Complicated, so bear with me. It involves *two* coins, one of which, the bad one, comes up heads only 9.5 percent of the time, tails 90.5 percent. The other coin, the good one, comes up heads 74.5 percent of the time, tails 25.5 percent. As in game S, you move up a stair if the coin you flip comes up heads and you move down one if it comes up tails.

But which coin do you flip? If the number of the stair you're on is a multiple of 3 (that is, . . . , –9, –6, –3, 0, 3, 6, 9,

12, . . .), you flip the bad coin. If the number of the stair you're on is not a multiple of 3, you flip the good coin. (Note: Changing these odd percentages and constraints may affect the game's outcome.)

Let's go through game C's dance steps. If you were on stair number 5, you would flip the good coin to determine your direction, whereas if you were on stair number 6, you would flip the bad coin. The same holds for the negatively numbered stairs. If you were on stair number –2 and playing game C, you would flip the good coin, whereas if you were on stair number –9, you would flip the bad coin.

Though less obviously so than in game S, game C is also a losing game. If you played it long enough, you would almost certainly reach the bottom of the staircase eventually. Game C is a losing game because the number of the stair you're on is a multiple of 3 more often than a third of the time and thus you must flip the bad coin more often than a third of the time. Take my word for this or read the next paragraph to get a better feel for why it is.

(Assume that you've just started playing game C. Since you're on stair number 0, and 0 is a multiple of 3, you would flip the bad coin, which lands heads with probability less than 10 percent, and you would very likely move down to stair number –1. Then, since –1 is not a multiple of 3, you would flip the good coin, which lands heads with probability almost 75 percent, and would probably move back up to stair 0. You may move up and down like this for a while. Occasionally, however, after the bad coin lands tails, the good coin, which lands tails almost 25 percent of the time, will land tails twice in succession, and you would move down to stair number –3, where the pattern will likely begin again. This latter downward pattern happens slightly more frequently (with probability $.905 \times .255 \times .255$) than does a rare head on the bad coin being followed by two heads on the good one (with probability .095 × .745 × .745) and your moving up three stairs as a consequence. So-called Markov chains are needed for a fuller analysis.)

So far, so what? Game S is simple and results in steady movement down the staircase to the bottom, and game C is complicated and also results in steady movement down the staircase to the bottom. Parrondo's fascinating discovery is that *if you play these two games in succession in random order (keeping your place on the staircase as you switch between games), you will steadily ascend to the top of the staircase.* Alternatively, if you play two games of S followed by two games of C followed by two games of S and so on, all the while keeping your place on the staircase as you switch between games, you will also steadily rise to the top of the staircase. (You might want to look up M. C. Escher's paradoxical drawing, "Ascending and Descending" for a nice visual analog to Parrondo's paradox.)

Standard stock-market investments cannot be modeled by games of this type, but variations of these games might conceivably give rise to counterintuitive investment strategies. The probabilities might be achieved, for example, by complicated combinations of various financial instruments (options, derivatives, and so on), but the decision which coin (which investment, that is) to flip (to make) in game C above would, it seems, have to depend upon something other than whether one's holdings were worth a multiple of $3.00 (or a multiple of $3,000.00). Perhaps the decision could depend in some way on the cross-correlation between a pair of stocks or turn on the value of some index being a multiple of 3.

If strategies like this could be made to work, they would yield what one day might be referred to as Parrondo profits.

Finally, let's consider a companion paradox of sorts that might be called "losing through winning" and that may help explain why companies often overpaid for small companies they were purchasing during the bubble in the late '90s. Professor Martin Shubik has regularly auctioned off $1 to students in his classes at Yale. The bidding takes place at 5¢ intervals, and the highest bidder gets the dollar, of course, but the second highest bidder is required to pay his bid as well. Thus, if the highest bid is 50¢ and you are second highest at 45¢, the leader stands to make 50¢ on the deal and you stand to lose 45¢ on it if bidding stops. You have an incentive to up your bid to at least 55¢, but after you've done so the other bidder has an even bigger incentive to raise his bid as well. In this way a one dollar bill can be successfully auctioned off for two, three, four, or more dollars.

If several companies are bidding on a small company and the cost of the preliminary legal, financial, and psychological efforts required to purchase the company are a reasonable fraction of the cost of the company, the situation is formally similar to Shubik's auction. One or more of the bidding companies might feel compelled to make an exorbitant preemptive offer to avoid the fate of the losing bidder on the $1. WorldCom's purchase of the web-hosting company Digex in 2000 for $6 billion was, I suspect, such an offer. John Sidgmore, the CEO who succeeded Bernie Ebbers, says that Digex was worth no more than $50 million, but that Ebbers was obsessed with beating out Global Crossing for the company.

The purchase is much more bizarre than Parrondo's paradox.

4 | Chance and Efficient Markets

*I*f the movement of stock prices is random or near-random, then the tools of technical analysis are nothing more than comforting blather giving one the illusion of control and the pleasure of a specialized jargon. They can prove especially attractive to those who tend to infuse random events with personal significance.

Even some social scientists don't seem to realize that if you search for a correlation between any two randomly selected attributes in a very large population, you will likely find some small but statistically significant association. It doesn't matter if the attributes are ethnicity and hip circumference, or (some measure of) anxiety and hair color, or perhaps the amount of sweet corn consumed annually and the number of mathematics courses taken. Despite the correlation's statistical significance (its unlikelihood of occurring by chance), it is probably not practically significant because of the presence of so many confounding variables. Furthermore, it will not necessarily support the (often ad hoc) story that accompanies it, the one purporting to explain why people who eat a lot of corn take more math. Superficially plausible tales are always available: Corn-eaters are more likely to be from the upper Midwest, where dropout rates are low.

Geniuses, Idiots, or Neither

Around stock market rises and declines, people are prone to devise just-so stories to satisfy various needs and concerns. During the bull markets of the '90s investors tended to see themselves as "perspicacious geniuses." During the more recent bear markets they've tended toward self-descriptions such as "benighted idiots."

My own family is not immune to the temptation to make up pat after-the-fact stories explaining past financial gains and losses. When I was a child, my grandfather would regale me with anecdotes about topics as disparate as his childhood in Greece, odd people he'd known, and the exploits of the Chicago White Sox and their feisty second baseman "Fox Nelson" (whose real name was Nelson Fox). My grandfather was voluble, funny, and opinionated. Only rarely and succinctly, however, did he refer to the financial reversal that shaped his later life. As a young and uneducated immigrant, he worked in restaurants and candy stores. Over the years he managed to buy up eight of the latter and two of the former. His candy stores required sugar, which led him eventually to speculate in sugar markets and—he was always a bit vague about the details—to place a big bet on several train cars full of sugar. He apparently put everything he had into the deal a few weeks before the sugar market crashed. Another version attributed his loss to underinsurance of the sugar shipment. In any case, he lost it all and never really recovered financially. I remember him saying ruefully, "Johnny, I would have been a very, very rich man. I should have known." The bare facts of the story registered with me then, but my recent less calamitous experience with WorldCom has made his pain more palpable.

This powerful natural proclivity to invest random events with meaning on many different levels makes us vulnerable to people who tell engaging stories about these events. In the

Rorschach blot that chance provides us, we often see what we want to see or what is pointed out to us by business prognosticators, distinguishable from carnival psychics only by the size of their fees. Confidence, whether justified or not, is convincing, especially when there aren't many "facts of the matter." This may be why market pundits seem so much more certain than, say, sports commentators, who are comparatively frank in acknowledging the huge role of chance.

Efficiency and Random Walks

The Efficient Market Hypothesis formally dates from the 1964 dissertation of Eugene Fama, the work of Nobel prize-winning economist Paul Samuelson, and others in the 1960s. Its pedigree, however, goes back much earlier, to a dissertation in 1900 by Louis Bachelier, a student of the great French mathematician Henri Poincare. The hypothesis maintains that at any given time, stock prices reflect all relevant information about the stock. In Fama's words: "In an efficient market, competition among the many intelligent participants leads to a situation where, at any point in time, actual prices of individual securities already reflect the effects of information based both on events that have already occurred and on events which, as of now, the market expects to take place in the future."

There are various versions of the hypothesis, depending on what information is assumed to be reflected in the stock price. The weakest form maintains that all information about past market prices is already reflected in the stock price. A consequence of this is that all of the rules and patterns of technical analysis discussed in chapter 3 are useless. A stronger version maintains that all publicly available information about a company is already reflected in its stock price. A consequence of this version is that the earnings, interest, and other elements of fundamental analysis discussed in chapter 5 are useless. The strongest version maintains that all information of all sorts is already reflected in the stock price. A consequence of this is that even inside information is useless.

It was probably this last, rather ludicrous version of the hypothesis that prompted the joke about the two efficient market theorists walking down the street: They spot a hundred dollar bill on the sidewalk and pass by it, reasoning that if it were real, it would have been picked up already. And of course there is the obligatory light-bulb joke. Question: How many efficient market theorists does it take to change a light bulb? Answer: None. If the light bulb needed changing the market would have already done it. Efficient market theorists tend to believe in passive investments such as broad-gauged index funds, which attempt to track a given market index such as the S&P 500. John Bogle, the crusading founder of Vanguard and presumably a believer in efficient markets, was the first to offer such a fund to the general investing public. His Vanguard 500 fund is unmanaged, offers broad diversification and very low fees, and generally beats the more expensive, managed funds. Investing in it does have a cost, however: One must give up the fantasy of a perspicacious gunslinger/investor outwitting the market.

And why do such theorists believe the market to be efficient? They point to a legion of investors of all sorts all seeking to make money by employing all sorts of strategies. These investors sniff out and pounce upon any tidbit of information even remotely relevant to a company's stock price, quickly driving it up or down. Through the actions of this investing horde the market rapidly responds to the new information, efficiently adjusting prices to reflect it. Opportunities to make an excess profit by utilizing technical rules or fundamental analyses, so the story continues, disappear before they can be fully exploited, and investors who pursue them will see their excess profits shrink to zero, especially after taking into account brokers' fees and other transaction costs. Once again, it's not that subscribers to technical or fundamental analysis won't make money; they generally will. They just won't make more than, say, the S&P 500.

(That exploitable opportunities tend to gradually disappear is a general phenomenon that occurs throughout economics and in a variety of fields. Consider an argument about baseball put forward by Steven Jay Gould in his book *Full House: The Spread of Excellence from Plato to Darwin.* The absence of .400 hitters in the years since Ted Williams hit .406 in 1941, he maintained, was not due to any decline in baseball ability but the reverse: a gradual increase in the athleticism of all players and a consequent decrease in the disparity between the worst and best players. When players are as physically gifted and well trained as they are now, the distribution of batting averages and earned run averages shows less variability. There are few "easy" pitchers for hitters and few "easy" hitters for pitchers. One result is that .400 averages are now very scarce. The athletic prowess of hitters and pitchers makes the "market" between them more efficient.)

There is, moreover, a close connection between the Efficient Market Hypothesis and the proposition that the movement of stock prices is random. If present stock prices already reflect all available information (that is, if the information is common knowledge in the sense of chapter 1), then future stock prices must be unpredictable. Any news that might be relevant in predicting a stock's future price has already been weighed and responded to by investors whose buying and selling have adjusted the present price to reflect the news. Oddly enough, as markets become more efficient, they tend to become less predictable. What will move stock prices in the future are truly new developments (or new shadings of old developments), news that is, by definition, impossible to anticipate. The conclusion is that in an efficient market, stock prices move up and down randomly. Evincing no memory of their past, they take what is commonly called a random walk, each step of which is independent of past steps. There is over time, however, an upward trend, as if the coin being flipped were slightly biased.

There is a story I've always liked that is relevant to the impossibility of anticipating new developments. It concerns a college student who completed a speed-reading course. He noted this fact in a letter to his mother. His mother responded with a long, chatty letter of her own in the middle of which she wrote, "Now that you've taken that speed-reading course, you've probably finished reading this letter by now."

Likewise, true scientific breakthroughs or applications, by definition, cannot be foreseen. It would be preposterous to have expected a newspaper headline in 1890 proclaiming "Only 15 Years Until Relativity." It is similarly foolhardy, the efficient market theorist reiterates, to predict changes in a company's business environment. To the extent these predictions reflect a consensus of opinion, they're already accounted for. To the extent that they don't, they're tantamount to forecasting coin flips.

Whatever your views on the subject, the arguments for an efficient market spelled out in Burton Malkiel's *A Random Walk Down Wall Street* and elsewhere can't be grossly wrong. After all, most mutual fund managers continue to generate average gains less than those of, say, the Vanguard Index 500 fund. (This has always seemed to me a rather scandalous fact.) There is other evidence for a *fairly* efficient market as well. There are few opportunities for risk-free money making or arbitrage, prices seem to adjust rapidly in response to news, and the autocorrelation of the stock prices from day to day, week to week, month to month, and year to year is small (albeit not zero). That is, if the market has done well (or poorly) over a given time period in the past, there is no strong tendency for it to do well (or poorly) during the next time period.

Nevertheless, in the last few years I have qualified my view of the Efficient Market Hypothesis and random-walk theory. One reason is the accounting scandals involving Enron, Adelphia, Global Crossing, Qwest, Tyco, WorldCom, Andersen, and many others from corporate America's Hall of Infamy, which make it hard to believe that available information about a stock always quickly becomes common knowledge.

Pennies and the Perception of Pattern

The *Wall Street Journal* has famously conducted a regular series of stock-picking contests between a rotating collection of stock analysts, whose selections are a result of their own studies, and dart-throwers, whose selections are determined randomly. Over many six-month trials, the pros' selections have performed marginally better than the darts' selections, but not overwhelmingly so, and there is some feeling that the pros' picks may influence others to buy the same stocks and hence drive up their price. Mutual funds, although less volatile than individual stocks, also display a disregard for analysts' pronouncements, often showing up in the top quarter of funds one year and in the bottom quarter the next.

Whether or not you believe in efficient markets and the random movement of stock prices, the huge element of chance present in the market cannot be denied. For this reason an examination of random behavior sheds light on many market phenomena. (So does study of a standard tome on probability such as that by Sheldon Ross.) Sources for such random behavior are penny stocks or, more accessible and more random, stocks of pennies, so let's imagine flipping a penny repeatedly and keeping track of the sequence of heads and tails. We'll assume the coin and the flip are fair (although, if we wish, the penny can be altered slightly to reflect the small upward bias of the market over time).

One odd and little-known fact about such a series of coin flips concerns the proportion of *time* that the number of heads exceeds the number of tails. It's seldom close to 50 percent!

To illustrate, imagine two contestants, Henry and Tommy, who bet that heads and tails respectively will be the outcome of a daily coin flip, a ritual that goes on for years. (Let's not ask why.) Henry is ahead on any given day if up to that day there have been more heads than tails, and Tommy is ahead if up to that day there have been more tails. The coin is fair, so they're equally likely to be in the lead, but one of them will probably be in the lead during most of their rather stultifying contest.

Stated numerically, the claim is that if there have been 1,000 coin flips, then it's considerably more probable that Henry (or Tommy) has been ahead more than, say, 96 percent of the time than that either one has been ahead between 48 percent and 52 percent of the time.

People find this result hard to believe. Many subscribe to the "gambler's fallacy" and believe that the coin's deviations from a 50–50 split between heads and tails are governed by a probabilistic rubber band: the greater the deviation, the greater the equalizing push toward an even split. But even if Henry were way ahead, with 525 heads to Tommy's 475 tails, his lead would be as likely to grow as to shrink. Likewise, a stock that's fallen on a truly random trajectory is as likely to fall further as it is to rise.

The rarity with which the lead switches sides in no way contradicts the fact that the proportion of heads approaches 1/2 as the number of flips increases. Nor does it contradict the phenomenon of regression to the mean. If Henry and Tommy were to start over and flip their penny another 1,000 times, it's quite likely that the number of heads would be smaller than 525.

Given the relative rarity with which Henry and Tommy overtake one another in their penny-flipping contest, it wouldn't be surprising if one of them came to be known as a "winner" and the other a "loser" despite their complete lack of control over the penny. If one professional stock picker outperformed another by a margin of 525 to 475, he might even be interviewed on Moneyline or profiled in *Fortune* magazine. Yet he might, like Henry or Tommy, owe his success to nothing more than getting "stuck" by chance on the up side of a 50–50 split.

But what about such stellar "value investors" as Warren Buffett? His phenomenal success, like that of Peter Lynch, John Neff, and others, is often cited as an argument against the market's randomness. This assumes, however, that Buffett's choices have no effect on the market. Originally no doubt they didn't, but now his selections themselves and his ability to create synergies among them can influence others. His performance is therefore a bit less remarkable than it first appears.

A different argument points to the near certainty of some stocks, funds, or analysts doing well over an extended period merely by chance. Of 1,000 stocks (or funds or analysts), for example, roughly 500 might be expected to outperform the market next year simply by chance, say by the flipping of a coin. Of these 500, roughly 250 might be expected to do well for a second year. And of these 250, roughly 125 might be expected to continue the pattern, doing well three years in a row simply by chance. Iterating in this way, we might reasonably expect there to be a stock (or fund or analyst) among the thousand that does well for ten consecutive years by chance alone. Once again, some in the business media are likely to go gaga over the performance.

The surprising length and frequency of consecutive runs of heads or tails is yet another lesson of penny flipping. If Henry and Tommy were to continue flipping pennies once a day, then there's a better-than-even chance that within about two months Henry will have won at least five flips in a row, as will Tommy. If they continue flipping for six years, there's a better-than-even chance that each will have won at least ten flips in a row.

When people are asked to write down a series of heads and tails that simulates a series of coin flips, they almost always fail to include enough runs of consecutive heads or consecutive tails. In particular, they fail to include any very long runs of heads or tails, and their series are thus easily distinguishable from a real series of coin flips.

But try telling people that long streaks are due to chance alone, whether the streak is a basketball player's shots, a stock analyst's picks, or a series of coin flips. The fact is that random events can frequently seem quite ordered.

To literally see this, take out a large piece of paper and partition it into little squares in a checkerboard pattern. Flip a coin repeatedly and color the squares white or black depending upon whether the coin lands heads or tails. After the checkerboard has been completely filled in, look it over and see if you can discern any patterns or clusters of similarly colored squares. Chances are you will, and if you felt the need to explain these patterns, you would invent a story that might sound superficially plausible or intriguing, but, given how the colors were determined, would necessarily be false.

The same illusion of pattern would result if you were to graph (with time on the horizontal axis) the results of the coin flips, up one unit for a head, down one for a tail. Some chartists and technicians would no doubt see "head and shoulders," "triple tops," or "ascending channels" patterns in these zigzag, up-and-down movements, and they would expatiate on their significance. (One difference between coin flips and models of random stock movements is that in the latter it is generally assumed that stocks move up or down not by a fixed amount per unit time, but by a fixed percentage.)

Leaving aside, once again, the question whether the market is perfectly efficient or whether stock movements follow a truly random walk, we can nevertheless say that phenomena that are truly random often appear almost indistinguishable from real-market behavior. This should, but probably won't, give pause to commentators who provide a neat post hoc explanation for every rally, every sell-off, and everything in between. Such commentators generally don't make remarks analogous to the observation that the penny happened by chance to land heads a few more times than it did tails. Instead they will refer to Tommy's profit-taking, Henry's increased confidence, labor problems in the copper mines, or countless other factors.

Because so much information is available—business pages, companies' annual reports, earnings expectations, alleged scandals, on-line sites, and commentary—something insightful-sounding can always be said. All we need do is filter the sea of numbers until we catch a plausible nugget of speculation. Like flipping a penny, doing so is a snap.

A Stock-Newsletter Scam

The accounting scandals involving WorldCom, Enron, and others derived from the data being selected, spun, and filtered. A scam I first discussed in my book *Innumeracy* derives instead from the *recipients* of the data being selected, spun, and filtered. It goes like this. Someone claiming to be the publisher of a stock newsletter rents a mailbox in a fancy neighborhood, has expensive stationery made up, and sends out letters to potential subscribers boasting of his sophisticated stock-picking software, financial acumen, and Wall Street connections. He writes also of his amazing track record, but notes that the recipients of his letters needn't take his word for it.

Assume you are one of these recipients and for the next six weeks you receive correct predictions about a certain common stock index. Would you subscribe to the newsletter? What if you received ten consecutive correct predictions?

Here's the scam. The newsletter publisher sends out 64,000 letters to potential subscribers. (Using email would save postage, but might appear to be a "spam scam" and hence be less credible.) To 32,000 of the recipients, he predicts the index in question will rise the following week and to the other 32,000, he predicts it will decline. No matter what happens to the index the next week, he will have made a correct prediction to 32,000 people. To 16,000 of them he sends another letter predicting a rise in the index for the following week, and to the other 16,000 he predicts a decline. Again, no matter what happens to the index the next week, he will have made correct predictions for two consecutive weeks to 16,000 people. To 8,000 of them he sends a third letter predicting a rise for the third week and to the other 8,000 he predicts a decline.

Focusing at each stage on the people to whom he's made only correct predictions and winnowing out the rest, he iterates this procedure a few more times until there are 1,000 people left to whom he's made six straight correct "predictions." To these he sends a different sort of follow-up letter, pointing out his successes and saying that they can continue to receive these oracular pronouncements if they pay the $1,000 subscription price to the newsletter. If they all pay, that's a million dollars for someone who need know nothing about stock, indices, trends, or dividends. If this is done knowingly, it is illegal. But what if it's done unknowingly by earnest, confident, and ignorant newsletter publishers? (Compare the faithhealer who takes credit for any accidental improvements.)

There is so much complexity in the market, there are so many different measures of success and ways to spin a story, that most people can manage to convince themselves that they've been, or are about to be, inordinately successful. If people are desperate enough, they'll manage to find some seeming order in random happenings.

Similar to the newsletter scam, but with a slightly different twist, is a story related to me by an acquaintance who described his father's business and its sad demise. He claimed that his father, years before, had run a large college-preparation service in a South American country whose identity I've forgotten. My friend's father advertised that he knew how to drastically improve applicants' chances of getting into the elite national university. Hinting at inside contacts and claiming knowledge of the various forms, deadlines, and procedures, he charged an exorbitant fee for his service, which he justified by offering a money-back guarantee to students who were not accepted.

One day, the secret of his business model came to light. All the material that prospective students had sent him over the years was found unopened in a trash dump. Upon investigation it turned out that he had simply been collecting the students' money (or rather their parents' money) and doing nothing for it. The trick was that his fees were so high and his marketing so focused that only the children of affluent parents subscribed to his service, and almost all of them were admitted to the university anyway. He refunded the fees of those few who were not admitted. He was also sent to prison for his efforts.

Are stock brokers in the same business as my acquaintance's father? Are stock analysts in the same business as the newsletter publisher? Not exactly, but there is scant evidence that they possess any unusual predictive powers. That's why I thought news stories in November 2002 recounting New York Attorney General Eliot Spitzer's criticism of *Institutional Investor* magazine's analyst awards were a tad superfluous. Spitzer noted that the stock-picking performances of most of the winning analysts were, in fact, quite mediocre. Maybe Donald Trump will hold a press conference pointing out that the country's top gamblers don't do particularly well at roulette.

Decimals and Other Changes

Like analysts and brokers, market makers (who make their money on the spread between the bid and the ask price for a stock) have received more than their share of criticism in recent years. One result has been a quiet reform that makes the market a bit more efficient. Wall Street's surrender to radical "decacrats" occurred a couple of years ago, courtesy of a Congressional mandate and a direct order from the Securities and Exchange Commission. Since then stock prices have been expressed in dollars and cents, and we no longer hear "profit-taking drove XYZ down 2 and 1/8" or "news of the deal sent PQR up 4 and 5/16."

Although there may be less romance associated with declines of 2.13 and rises of 4.31, decimalization makes sense for a number of reasons. The first is that price rises and declines are immediately comparable since we no longer must perform the tiresome arithmetic of, say, dividing 11 by 16. Mentally calculating the difference between two decimals generally requires less time than subtracting 3 5/8 from 5 3/16. Another benefit is global uniformity of pricing, as American securities are now denominated in the same decimal units as those in the rest of the world. Foreign securities no longer need to be rounded to the nearest multiple of 1/16, a perverse arithmetical act if there ever was one.

More importantly, the common spread between the bid and ask prices has shrunk. Once generally 1/16 (.0625, that is), the spread in many cases has become .01 and, by so shriveling, will save investors billions of dollars over the years. Market makers aside, most investors applaud this consequence of decimalization.

The last reason for cheering the change is more mathematical. There is a sense in which the old system of halves, quarters, eighths, and sixteenths is more natural than decimals. It is, after all, only a slightly disguised binary system, based on powers of 2 (2, 4, 8, 16) rather than powers of 10. It doesn't inherit any of the prestige of the binary system, however, because it awkwardly combines the base 2 fractional part of a stock price with the base 10 whole-number part.

Thus it is that Ten extends its imperial reach to Wall Street. From the biblical Commandments to David Letterman's lists, the number 10 is ubiquitous. Not unrelated to the perennial yearning for the simplicity of the metric system, 10 envy has also come to be associated with rationality and efficiency. It is thus fitting that all stocks are now expressed in decimals. Still, I suspect that many market veterans miss those pesky fractions and their role in stories of past killings and baths. Except for generation X-ers (Roman numeral ten-ers), many others will too. Anyway, that's my two cents (.02, 1/50th) worth on the subject.

The replacement of marks, francs, drachmas, and other European currencies by euros on stock exchanges and in stores is another progressive step that nevertheless rouses a touch of nostalgia. The coins and bills from my past travels that are scattered about in drawers are suddenly out of work and will never see the inside of a wallet again.

Yet another vast change in trading practices is the greater self-reliance among investors. Despite the faulty accounting that initially disguised their sickly returns, the ladies of Beardstown, Illinois, helped popularize investment clubs. Even more significant in this regard is the advent of effortless online trading, which has further hastened the decline of the traditional broker. The ease with which I clicked on simple icons to buy and sell (specifically sell reasonably performing funds and buy more WCOM shares) was always a little frightening, and I sometimes felt as if there were a loaded gun on my desk. Some studies have linked online trading and day trading to increased volatility in the late '90s, although it's not clear that they remain factors in the '00s.

What's undeniable is that buying and selling online remains easy, so easy that I think it might not be a bad idea were small pictures of real-world items to pop up before every stock purchase or sale as a reminder of the approximate value of what's being traded. If your transaction were for $35,000, a luxury car might appear; if it were for $100,000, a small cottage; and if it were for a penny stock, a candy bar. Investors can now check stock quotations, the size and the number of the bids and the asks, and megabytes of other figures on so-called level-two screens available in (almost) real-time on their personal computers. Millions of little desktop brokerages! Unfortunately, librarian Jesse Sherra's paraphrase of Coleridge often seems apt: Data, data everywhere, but not a thought to think.

Benford's Law and Looking Out for Number One

I mentioned that people find it very difficult to simulate a series of coin flips. Are there other human disabilities that might allow someone to look at a company's books, say Enron's or WorldCom's, and determine whether or not they had been cooked? There may have been, and the mathematical principle involved is easily stated, but counterintuitive.

Benford's Law states that in a wide variety of circumstances, numbers—as diverse as the drainage areas of rivers, physical properties of chemicals, populations of small towns, figures in a newspaper or magazine, and the half-lives of radioactive atoms—have "1" as their first non-zero digit disproportionately often. Specifically, they begin with "1" about 30 percent of the time, with "2" about 18 percent of the time, with "3" about 12.5 percent, and with larger digits progressively less often. Less than 5 percent of the numbers in these circumstances begin with the digit "9." Note that this is in stark contrast to many other situations where each of the digits has an equal chance of appearing.

Benford's Law goes back one hundred years to the astronomer Simon Newcomb (note the letters WCOM in his name), who noticed that books of logarithm tables were much dirtier near the front, indicating that people more frequently looked up numbers with a low first digit. This odd phenomenon remained a little-known curiosity until it was rediscovered in 1938 by the physicist Frank Benford. It wasn't until 1996, however, that Ted Hill, a mathematician at Georgia Tech, established what sort of situations generate numbers in accord with Benford's Law. Then a mathematically inclined accountant named Mark Nigrini generated considerable buzz when he noted that Benford's Law could be used to catch fraud in income tax returns and other accounting documents.

The following example suggests why collections of numbers governed by Benford's Law arise so frequently:

Imagine that you deposit $1,000 in a bank at 10 percent compound interest per year. Next year you'll have $1,100, the year after that $1,210, then $1,331, and so on. (Compounding is discussed further in chapter 5.) The first digit of your account balance remains a "1" for a long time. When your account grows to over $2,000, the first digit will remain a "2" for a shorter period. And when your deposit finally grows to over $9,000, the 10 percent growth will result in more than $10,000 in your account the following year and a long return to "1" as the first digit. If you record your account balance each year for many years, these numbers will thus obey Benford's Law.

The law is also "scale-invariant" in that the dimensions of the numbers don't matter. If you expressed your $1,000 in euros or pounds (or the now defunct francs or marks) and watched it grow at 10 percent per year, about 30 percent of the yearly values would begin with a "1," about 18 percent with a "2," and so on.

More generally, Hill showed that such collections of numbers arise whenever we have what he calls a "distribution of distributions," a random collection of random samples of data. Big, motley collections of numbers will follow Benford's Law.

This brings us back to Enron, WorldCom, accounting, and Mark Nigrini, who reasoned that the numbers on accounting forms, which often come from a variety of company operations and a variety of sources, should be governed by Benford's Law. That is, these numbers should begin disproportionately with the digit "1," and progressively less often with bigger digits, and if they don't, that is a sign that the books have been cooked. When people fake plausible-seeming numbers, they generally use more "5s" and "6s" as initial digits, for example, than Benford's Law would predict.

Nigrini's work has been well publicized and has surely been noted by accountants and by prosecutors. Whether the Enron, WorldCom, and Andersen people have heard of it is unknown, but investigators might want to check if the distribution of leading digits in the Enron documents accords with Benford's Law. Such checks are not foolproof and sometimes lead to false-positive results, but they provide an extra tool that might be useful in certain situations.

It would be amusing if, in looking out for number one, the culprits forgot to look out for their "1s." Imagine the Andersen accountants muttering anxiously that there weren't enough leading "1s" on the documents they were feeding into the shredders. A 1-derful fantasy!

The Numbers Man—A Screen Treatment

An astonishing amount of attention has been paid recently to fictional and narrative treatments of mathematical topics. The movies *Good Will Hunting*, *Pi*, and *The Croupier* come to mind; so do plays such as *Copenhagen*, *Arcadia*, and *The Proof*, the two biographies of Paul Erdos, *A Beautiful Mind*, the biography of John Nash (with its accompanying Academy Award-winning movie), TV specials on Fermat's Last Theorem, and other mathematical topics, as well as countless books on popular mathematics and mathematicians. The plays and movies, in particular, prompted me to expand the idea in the stock-newsletter scam discussed above (I changed the focus, however, from stocks to sports) into a sort of abbreviated screen treatment that highlights the relevant mathematics a bit more than has been the case in the productions just cited. Yet another instance of what columnist Charles Krauthammer has dubbed "Disturbed Nerd Chic," the treatment might even be developed into an intriguing and amusing film. In fact, I rate it a "strong buy" for any studio executive or independent filmmaker.

Rough Idea: Math nerd runs a clever sports-betting scam and accidentally nets an innumerate mobster.

Act One

Louis is a short, lecherous, somewhat nerdy man who dropped out of math graduate school about ten years ago (in the late '80s) and now works at home as a technical consultant. He looks and acts a bit like the young Woody Allen. He's playing cards with his pre-teenage kids and has just finished telling them a funny story. His kids are smart and they ask him how it is that he always knows the right story to tell. His wife, Marie, is uninterested. True to form, he begins telling them the Leo Rosten story about the famous rabbi who was asked by an admiring student how it was that the rabbi always had a perfect parable for any subject. Louis pauses to make sure they see the relevance.

When they smile and his wife rolls her eyes again, he continues. He tells them that the rabbi replied to his students with a parable. It was about a recruiter in the Tsar's army who was riding through a small town and noticed dozens of chalked circular targets on the side of a barn, each with a bullet hole through the bull's-eye. The recruiter was impressed and asked a neighbor who this perfect shooter might be. The neighbor responded, "Oh that's Shepsel, the shoemaker's son. He's a little peculiar." The enthusiastic recruiter was undeterred until the neighbor added, "You see, first Shepsel shoots and *then* he draws the chalk circles around the bullet hole." The rabbi grinned. "That's the way it is with me. I don't look for a parable to fit the subject. I introduce only subjects for which I have parables."

Louis and his kids laugh until a distracted, stricken look crosses his face. Closing the book, Louis hurries his kids off to bed, interrupts Marie's prattling about her new pearl necklace and her Main Line parents' nasty neighbors, distractedly bids her good night, and retreats to his study where he starts scribbling, making calls, and performing calculations. The next day he stops by the bank and the post office and a stationery store, does some research online, and then has a long discussion with his friend, a sportswriter on the local suburban New Jersey newspaper. The conversation revolves around the names, addresses, and intelligence of big sports bettors around the country.

The idea for a lucrative con game has taken shape in his mind. For the next several days he sends letters and emails to many thousands of known sports bettors "predicting" the outcome of a certain sporting event. His wife is uncomprehending when Louis mumbles that, Shepsel-like, he can't lose since whatever happens in the sporting event, his prediction is bound to be right for half the bettors. The reason, it will turn out, is that to half of these people he predicts a certain team will win, and to the other half he predicts that it will lose.

Tall, blond, plain, and dim-witted, Marie is left wondering what exactly her sneaky husband is up to now. She finds the new postage meter behind the computer, notes the increasingly frequent secret telephone calls, and nags him about their worsening financial and marital situation. He replies that she doesn't really need three closets full of clothes and a small fortune of jewelry when she spends all her time watching soaps and puts her off with some mathematical mumbo-jumbo about demographic research and new statistical techniques. She still doesn't follow, but she is mollified by his promise that his mysterious endeavor will end up being lucrative.

They go out to eat to celebrate and Louis, intense and cad-like as always, talks up genetically modified food and tells the cute waitress that he wants to order whatever item on the menu has the most artificial ingredients. Much to Marie's chagrin, he then involves the waitress in a classic mathematical trick by asking her to examine his three cards, one black on both sides, one red on both sides, and one black on one side and red on the other. He asks her for her cap, drops the cards into it, and tells her to pick a card, but only to look at one side of it. The side is red, and Louis notes that the card she picked couldn't possibly be the card that was black on both sides, and therefore it must be one of the other two cards—the red-red card or the red-black card. He guesses that it's the red-red card and offers to double her 15 percent tip if it's the red-black card and stiff her if it's the red-red card. He looks at Marie for approbation that is not forthcoming. The waitress accepts and loses.

Tone-deaf to Marie's discomfort, Louis thinks he's making amends with her by explaining the trick. She is less than enthralled. He tells her that it's not an even bet even though at first glance it appears to be one. There are, after all, two cards it could be, and he bet on one, and the waitress bet on the other. The rub is, he gleefully runs on with his mouth full, there are two ways he can win and only one way the waitress can win. The visible side of the card the waitress picked could be the red side of the red-black card, in which case she wins, or it could be one side of the red-red card, in which case he wins, or it could be the other side of the red-red card, in which case he also wins. His chances of winning are thus 2/3, he concludes exultantly, and the average tip he gives is reduced by a third. Marie yawns and checks her Rolex. He breaks to go to the men's room where he calls his girlfriend May Lee to apologize for some vague indiscretion.

The next week he explains the sports-betting con to May Lee, who looks a bit like Lucy Liu and is considerably smarter than Marie and even more materialistic. They're in her apartment. She is interested in the con and asks clarifying questions. He enthuses to her that he needs her secretarial help. He's sending out more letters and making a second prediction in them, but this time just to the half of the people to whom he sent a correct first prediction; the other half he plans to ignore. To half of this smaller group, he will predict a win in a second sporting event, to the other half a loss. Again for half of this group his prediction is going to be right, and so for one-fourth of the original group he's going to be right two times in a row. "And to this one-fourth of the bettors?" she asks knowingly and excitedly. A mathematico-sexual tension develops.

He smiles rakishly and continues. To half of this one-fourth he will predict a win the following week, to the other half a loss; he again will ignore those to whom he's made an incorrect prediction. Once again he will be right—this time for the third straight time—although for only one-eighth of the original population. May Lee helps with the mailings as he continues this process, focusing only on those to whom he's made correct predictions and winnowing out those to whom he's made incorrect ones. There is a sex scene amid all the letters, and they joke about winning whether the teams in question do or not, whether the predictions are right or wrong. Whether up or down, they'll be happy.

As the mailings go on, so does his other life as a bored consultant, cyber-surfer, and ardent sports fan. He continues to extend his string of successful predictions to a smaller and smaller group of people until finally with great anticipation he sends a letter to the small group of people who are left. In it he points to his impressive string of successes and requests a substantial payment to keep these valuable and seemingly oracular "predictions" coming.

Act Two

He receives many payments and makes a further prediction. Again he's right for half of the remaining people and drops the half for which he's wrong. He asks the former group for even more money for another prediction, receives it, and continues. Things improve with Marie and with May Lee as the money rolls in and Louis realizes his plan is working even better than he expected. He takes his kids and, in turn, each woman to sports events or to Atlantic City, where he comments smugly on the losers who, unlike him, bet on iffy propositions. When Marie worries aloud about shark attacks off the beach, Louis tells her that more Americans die from falling airplane parts each year than from shark attacks. He makes similar pronouncements throughout the trip.

He plays a little blackjack and counts cards while doing so. He complains that it requires too much low-level concentration and that, unless one has a lot of money already, the rate at which one makes money is so slow and uneven that one might as well get a job. Still, he goes on, it's the only game where a strategy exists for winning. All the other games are for mentally flabby losers. He goes to one of the casino restaurants where he shows his kids the waitress tip-cheating game. They think it's great.

Back home in suburban New Jersey again, the sports-betting con resumes. Now there are only a few people left among the original thousands of sports-bettors. One of them, a rough underworld type named Otto, tracks him down, follows him to the parking lot of the basketball arena, and, politely at first and then more and more insistently, demands a prediction on an upcoming game on which he plans to bet a lot of money. Louis dismisses him and Otto, who looks a little like Stephen Segal, promptly orders him into his car at gunpoint and threatens to harm his family. He knows where they live.

Not understanding how he could be the recipient of so many consecutive correct predictions, Otto doesn't believe Louis's protestations that this is a con game. Louis makes some mathematical points in an effort to convince Otto of the possible falsity of any particular prediction. But no matter how he tries, he can't quite convince Otto of the fact that there will always be some people who receive many consecutive correct predictions by chance alone.

Marooned in Otto's basement, the math-nerd scam artist and the bald muscled extortionist are a study in contrasts. They speak different languages and have different frames of reference. Otto claims, for example, that every bet is more or less a 50–50 proposition because you either win or lose. Louis talks of his basketball buddies Lewis Carroll and Bertrand Russell and the names go over Otto's head, of course. Oddly, they have similar attitudes toward women and money and also share an interest in cards, which they play to while away the time. Otto proudly shows off his riffle shuffle that he claims completely mixes the cards, while Louis prefers solitaire and silently scoffs at Otto's lottery expenditures and gambling misconceptions. When they forget why they're there they get along well enough, although now and then Otto renews his threats and Louis renews his disavowal of any special sports knowledge and his plea to go home.

Finally granting that he might receive an incorrect prediction occasionally, Otto still insists that Louis give him his take on who's going to win an upcoming football game. In addition to not being too bright, Otto, it appears, is in serious debt. Under extreme duress (with a gun to his head), Louis makes a prediction that happens to be right, and Otto, desperate and still convinced that he is in control of a money tree, now wants to bet funds borrowed from his gambling associates on Louis's next prediction.

Act Three

Louis at last convinces Otto to let him go home and do research for his next big sports prediction. He and May Lee, whose need for money, baubles, and clothes has all along provided the impetus for the scam, discuss his predicament and realize they must exploit Otto's only weaknesses, his stupidity and gullibility, and his only intellectual interests, money and playing cards.

Both go over to Otto's apartment. Otto is charmed by May Lee, who flirts with him and offers him a deal. She wordlessly takes two decks of cards from her purse and asks Otto to shuffle each of them. Otto is pleased to show off to a more appreciative audience. She then gives him one of the decks and asks him to turn over one card at a time as she, keeping pace with him, does the same thing with the other deck. May Lee asks, what does he think is the likelihood that the cards they turn over will ever match, denomination and suit exactly the same? He scoffs but is entranced by May Lee and is amazed when after a tense minute or so that is exactly what happens. She explains that it will happen more often than not and perhaps he can use this fact to make some money. After all, Louis is a mathematical genius and he's proved that it will. Louis smiles proudly.

Otto is puzzled. May Lee tells Otto again that the sports-betting was a scam and that he's more likely to make money with the card tricks that Louis can teach him. Louis steps forward with the same two decks, which he's now arranged so that the cards in each deck alternate colors. In one, it's red-black, red-black, red-black . . . . In the other it's black-red, black-red, black-red . . . . He gives the two decks to Otto and challenges him to do one of his great riffle shuffles of one deck into the other so that the cards will be mixed. Otto does and arrogantly announces that the cards are completely mixed now, whereupon Louis takes the combined two decks, puts them behind his back, pretends to be manipulating them, and brings forward two cards, one black and one red. So, Otto asks? Louis brings forth two more cards, one of each color, and then he does this again and again. I really shuffled them, Otto observes. How'd you do that? Louis explains that it involves no skill; the cards no longer alternate color in the combined deck, but any two from the top on down are always of different color.

There is a collage scene in which Louis explains various card tricks to Otto and the ways in which they can be exploited to make money. There's always some order, some deviation from randomness, that a card man like you can use to get rich, Louis says to Otto. He even explains to him how he avoids paying waitresses tips. The deal, of course, is that Otto releases them, understanding, vaguely at least, how the betting scheme works and, more precisely, how the new card tricks do. Louis promises a one-day crash course on how to exploit the tricks for money.

In the last scene Louis is seen working the same scam but this time with predictions about the movements of a stock market index. Since he doesn't want any more Ottos, but a higher class clientele, he's redefined himself as the publisher of a stock newsletter. The house he's in is more sumptuous and May Lee, to whom he's now married, bustles about in an expensive suit as Louis plays cards with his slightly older children, occasionally doodling little bull's-eyes and targets on an envelope. He excuses himself and goes to his study to make a secret telephone call to an apartment on Central Park West that he's just purchased for his new mistress.

5 | Value Investing and Fundamental Analysis

I was especially smitten with WorldCom's critical Internet division, UUNet. The Internet wasn't going away, and so, I thought, neither was UUNet or WorldCom. During this time of enchantment my sensible wife would say "UUNet, UUNet" and roll her pretty eyes to mock my rhapsodizing about WorldCom's global IP network and related capabilities. The repetition of the word gradually acquired a more general anti-Pollyannish meaning as well. "Maybe the bill is so exorbitant because the plumber ran into something he didn't expect." "Yeah, sure. UUNet, UUNet."

"Smitten," "rhapsodizing," and "Pollyanna" are not words that come naturally to mind when discussing value investing, a major approach to the market that uses the tools of so-called fundamental analysis. Often associated with Warren Buffett's gimlet-eyed no-nonsense approach to trading, fundamental analysis is described by some as the best, most sober strategy for investors to follow. Had I paid more attention to WorldCom fundamentals, particularly its $30 billion in debt, and less attention to WorldCom fairy tales, particularly its bright future role as a "dumb" network (better not to ask), I would no doubt have fared better. In the stock market's enduring tug-of-war between statistics and stories, fundamental analysis is generally on the side of the numbers.

Still, fundamental analysis has always seemed to me slightly at odds with the general ethic of the market, which is based on hope, dreams, vision, and a certain monetarily tinted yet genuine romanticism. I cite no studies or statistics to back up this contention, only my understanding of the investors I've known or read about and perhaps my own infatuation, quite atypical for this numbers man, with WorldCom.

Fundamentals are to investing what (stereotypically) marriage is to romance or what vegetables are to eating—healthful, but not always exciting. Some understanding of them, however, is essential for any investor and, to an extent, for any intelligent citizen. Everybody's heard of people who refrain from buying a house, for example, because of the amount they would have paid in interest over the years. ("Oh my, don't get a mortgage. You'll end up paying four times as much.") Also common are lottery players who insist that the worth of their possible winnings is really the advertised one million dollars. ("In only 20 years, I'll have that million.") And there are many investors who doubt that the opaque pronouncements of Alan Greenspan have anything to do with the stock or bond markets.

These and similar beliefs stem from misconceptions about compound interest, the bedrock of mathematical finance, which is in turn the foundation of fundamental analysis.

e is the Root of All Money

Speaking of bedrocks and foundations, I claim that e is the root of all money. That's e as in e^x as in exponential growth as in compound interest. An old adage (probably due to an old banker) has it that those who understand compound interest are more likely to collect it, those who don't more likely to pay it. Indeed the formula for such growth is the basis for most financial calculations. Happily, the derivation of a related but simpler formula depends only on understanding percentages, powers, and multiplication—on knowing, for example, that 15 percent of 300 is .15 × 300 (or 300 × .15) and that 15 percent of 15 percent of 300 is 300 × $(.15)^2$.

With these mathematical prerequisites stated, let's begin the tutorial and assume that you deposit $1,429.73 into a bank account paying 6.9 percent interest compounded annually. No, let's bow to the great Rotundia, god of round numbers, and assume instead that you deposit $1,000 at 10 percent. After one year, you'll have 110 percent of your original deposit—$1,100. That is, you'll have 1,000 × 1.10 dollars in your account. (The analysis is the same if you buy $1,000 worth of some stock and it returns 10 percent annually.)

Looking ahead, observe that after two years you'll have 110 percent of your first-year balance—$1,210. That is, you'll have ($1,000 × 1.10) × 1.10. Equivalently, that is $1,000 × 1.10^2. Note that the exponent is 2.

After three years you'll have 110 percent of your second-year balance—$1,331. That is, you'll have ($1,000 × 1.10^2) × 1.10. Equivalently, that is $1,000 × 1.10^3. Note the exponent is 3 this time.

The drill should be clear now. After four years you'll have 110 percent of your third-year balance—$1,464.10. That is, you'll have ($1,000 × 1.10^3) × 1.10. Equivalently, that is $1,000 × 1.10^4. Once again, note the exponent is 4.

Let me interrupt this relentless exposition with the story of a professor of mine long ago who, beginning at the left side of a very long blackboard in a large lecture hall, started writing 1 + 1/1! + 1/2! + 1/3! + 1/4! + 1/5! . . . . (Incidentally the expression 5! is read 5 factorial, not 5 with an exclamatory flair, and it is equal to 5 × 4 × 3 × 2 × 1. For any whole number N, N! is defined similarly.) My fellow students initially laughed as this professor, slowly and seemingly in a trance, kept on adding terms to this series. The laughter died out, however, by the time he reached the middle of the board and was writing 1/44! + 1/45! + ... I liked him and remember a feeling of alarm as I saw him continue his senseless repetitions. When he came to the end of the board at 1/83!, he turned and faced the class. His hand shook, the chalk dropped to the floor, and he left the room and never returned.

Mindful thereafter of the risks of too many illustrative repetitions, especially when I'm standing at a blackboard in a classroom, I'll end my example with the fourth year and simply note that the amount of money in your account after t years will be $1,000 × 1.10^t$. More generally, if you deposit P dollars into an account earning r percent interest annually, it will be worth A dollars after t years, where $A = P(1 + r)^t$, the promised formula describing exponential growth of money.

You can adjust the formula for interest compounded semi-annually or monthly or daily. If money is compounded four times per year, for example, then the amount you'll have after t years is given by $A = P(1 + r/4)^{4t}$. (The quarterly interest rate is r/4, one-fourth the annual rate of r, and the number of compoundings in t years is 4t, four per year for t years.)

If you compound very frequently (say n times per year for a large number n), the formula $A = P(1 + r/n)^{nt}$ can be mathematically massaged and rewritten as $A = Pe^{rt}$, where e, approximately 2.718, is the base of the natural logarithm. This variant of the formula is used for continuous compounding (and is, of course, the source of my comment that e is the root of all money).

The number e plays a critical role in higher mathematics, best exemplified perhaps by the formula $e^{\pi i} + 1 = 0$, which packs the five arguably most important constants in mathematics into a single equation. The number e also arises if we're simply choosing numbers between 0 and 1 at random. If we (or, more likely, our computer) pick these numbers until their sum exceeds 1, the average number of picks we'd need would be e, about 2.718. The ubiquitous e also happens to equal $1 + 1/1! + 1/2! + 1/3! + 1/4! + \ldots$, the same expression my professor was writing on the board many years ago. (Inspired by a remark by stock speculator Ivan Boesky, Gordon Gecko in the 1987 movie *Wall Street* stated, "Greed is good." He misspoke. He intended to say, "e is good.")

Many of the formulas useful in finance are consequences of these two formulas: $A = P(1 + r)^t$ for annual compounding and, for continuous compounding, $A = Pe^{rt}$. To illustrate how they're used, note that if you deposit $5,000 and it's compounded annually for 12 years at 8 percent, it will be worth $5,000(1.08)^{12}$ or $12,590.85. If this same $5,000 is compounded continuously, it will be worth $5,000e^{(.08 \times 12)}$ or $13,058.48.

Using this interest rate and time interval, we can say that the future value of the present $5,000 is $12,590.85 and that the present value of the future $12,590.85 is $5,000. (If the compounding is continuous, substitute $13,058.48 in the previous sentence.) The "present value" of a certain amount of future money is the amount we would have to deposit now so that the deposit would grow to the requisite amount in the allotted time. Alternatively stated (repetition may be an occupational hazard of professors; so may self-reference), the idea is that given an interest rate of 8 percent, you should be indifferent between receiving $5,000 now (the present value) and receiving something near $13,000 (the future value) in twelve years.

And just as "George is taller than Martha" and "Martha is shorter than George" are different ways to state the same relation, the interest formulas may be written to emphasize either present value, P, or future value, A. Instead of $A = P(1 + r)^t$, we can write $P = A/(1 + r)^t$, and instead of $A = Pe^{rt}$, we can write $P = A/e^{rt}$. Thus, if the interest rate is 12 percent, the present value of \$50,000 five years hence is given by P = \$50,000/$(1.12)^5$ or \$28,371.34. This amount, \$28,371.34, if deposited at 12 percent compounded annually for five years, has a future value of \$50,000.

One consequence of these formulas is that the "doubling time," the time it takes for a sum of money to double in value, is given by the so-called rule of 72: divide 72 by 100 times the interest rate. Thus, if you can get an 8 percent (.08) rate, it will take you 72/8 or nine years for a sum of money to double, eighteen years for it to quadruple, and twenty-seven years for it to grow to eight times its original size. If you're lucky enough to have an investment that earns 14 percent, your money will double in a little more than five years (since 72/14 is a bit more than 5) and quadruple in a bit over ten years. For continuous compounding, you use 70 rather than 72.

These formulas can also be used to determine the so-called internal rate of return and to define other financial concepts. They provide as well the muscle behind common pleas to young people to begin saving and investing early in life if they wish to become the "millionaire next door." (They don't, however, tell the millionaire next door what he should do with his wealth.)

The Fundamentalists' Creed: You Get What You Pay For

The notion of present value is crucial to understanding the fundamentalists' approach to stock valuation. It should also be important to lottery players, mortgagors, and advertisers. That the present value of money in the future is less than its nominal value explains why a nominal \$1,000,000 award for winning a lottery—say \$50,000 per year at the end of each of the next twenty years—is worth considerably less than $1,000,000. If the interest rate is 10 percent annually, for example, the $1,000,000 has a present value of only about $426,000. You can obtain this value from tables, from financial calculators, or directly from the formulas above (supplemented by a formula for the sum of a so-called geometric series).

The process of determining the present value of future money is often referred to as "discounting." Discounting is important because, once you assume an interest rate, it allows you to compare amounts of money received at different times. You can also use it to evaluate the present or future value of an income stream—different amounts of money coming into or going out of a bank or investment account on different dates. You simply "slide" the amounts forward or backward in time by multiplying or dividing by the appropriate power of $(1 + r)$. This is done, for example, when you need to figure out a payment sufficient to pay off a mortgage in a specified amount of time or want to know how much to save each month to have sufficient funds for a child's college education when he or she turns eighteen.

Discounting is also essential to defining what is often called a stock's fundamental value. The stock's price, say investing fundamentalists (fortunately not the sort who wish to impose their moral certitudes on others), should be roughly equal to the discounted stream of dividends you can expect to receive from holding onto it indefinitely. If the stock does not pay dividends or if you plan on selling it and thereby realizing capital gains, its price should be roughly equal to the discounted value of the price you can reasonably expect to receive when you sell the stock plus the discounted value of any dividends. It's probably safe to say that most stock prices are higher than this. During the 1990 boom years, investors were much more concerned with capital gains than they were with dividends. To reverse this trend, finance professor Jeremy Siegel, author of *Stocks for the Long Run*, and two of his colleagues recently proposed eliminating the corporate dividend tax and making dividends deductible.

The bottom line of bottom-line investing is that you should pay for a stock an amount equal to (or no more than) the present value of all future gains from it. Although this sounds very hard-headed and far removed from psychological considerations, it is not. The discounting of future dividends and the future stock price is dependent on your estimate of future interest rates, dividend policies, and a host of other uncertain quantities, and calling them fundamentals does not make them immune to emotional and cognitive distortion. The tango of exuberance and despair can and does affect estimates of stock's fundamental value. As the economist Robert Shiller has long argued quite persuasively, however, the fundamentals of a stock don't change nearly as much or as rapidly as its price.

Ponzi and the Irrational
Discounting of the Future

Before returning to other applications of these financial notions, it may be helpful to take a respite and examine an extreme case of undervaluing the future: pyramids, Ponzi schemes, and chain letters. These differ in their details and colorful storylines. A recent example in California took the form of all-women dinner parties whose new members contributed cash appetizers. Whatever their outward appearance, however, almost all these scams involve collecting money from an initial group of "investors" by promising them quick and extraordinary returns. The returns come from money contributed by a larger group of people. A still larger group of people contributes to both of the smaller earlier groups.

This burgeoning process continues for a while. But the number of people needed to keep the pyramid growing and the money coming in increases exponentially and soon becomes difficult to maintain. People drop out, and the easy marks become scarcer. Participants usually lack a feel for how many people are required to keep the scheme going. If each of the initial group of ten recruits ten more people, for example, the secondary group numbers 100. If each of these 100 recruit ten people, the tertiary group numbers 1,000. Later groups number 10,000, then 100,000, then 1,000,000. The system collapses under its own weight when enough new people can no longer be found. If you enter the scheme early, however, you can make extraordinarily quick returns (or could if such schemes were not illegal).

The logic of pyramid schemes is clear, but people generally worry only about what happens one or two steps ahead and anticipate being able to get out before a collapse. It's not irrational to get involved if you are confident of recruiting a "bigger sucker" to replace you. Some would say that the dot-coms' meteoric stock price rises in the late '90s and their subsequent precipitous declines in 2000 and 2001 were attenuated versions of the same general sort of scam. Get in on the initial public offering, hold on as the stock rockets upward, and jump off before it plummets.

Although not a dot-com, WorldCom achieved its all-too-fleeting dominance by buying up, often for absurdly inflated prices, many companies that were (and a good number that weren't). MCI, MFS, ANS Communication, CAI Wireless, Rhythms, Wireless One, Prime One Cable, Digex, and dozens more companies were acquired by Bernie Ebbers, a pied piper whose song seemed to consist of only one entrancing and repetitive note: acquire, acquire, acquire. The regular drumbeat of WorldCom acquisitions had the hypnotic quality of the tinkling bells that accompany the tiniest wins at casino slot machines. As the stock began its slow descent, I'd check the business news every morning and was tranquilized by news of yet another purchase, web hosting agreement, or extension of services.

While corporate venality and fraud played a role in (some of) their falls, the collapses of the dot-coms and WorldCom were not the brainchilds of con artists. Even when entrepreneurs and investors recognized the bubble for what it was, most figured incorrectly that they'd be able to find a chair when the mania-inducing IPO/acquisition music stopped. Alas, the journey from "have-lots" to "have-nots" was all too frequently by way of "have-dots."

Maybe our genes are to blame. (They always seem to get the rap.) Natural selection probably favors organisms that respond to local or near-term events and ignore distant or future ones, which are discounted in somewhat the same way that future money is. Even the ravaging of the environment may be seen as a kind of global Ponzi scheme, the early "investors" doing well, later ones less well, until a catastrophe wipes out all gains.

A quite different illustration of our short-sightedness comes courtesy of Robert Louis Stevenson's "The Imp in the Bottle." The story tells of a genie in a bottle able and willing to satisfy your every romantic whim and financial desire. You're offered the opportunity to buy this bottle and its amazing denizen at a price of your choice. There is a serious limitation, however. When you've finished with the bottle, you have to sell it to someone else at a price strictly less than what you paid for it. If you don't sell it to someone for a lower price, you will lose everything and will suffer excruciating and unrelenting torment. What would you pay for such a bottle?

Certainly you wouldn't pay 1 cent because then you wouldn't be able to sell it for a lower price. You wouldn't pay

2 cents for it either since no one would buy it from you for 1 cent since everyone knows that it must be sold for a price less than the price at which it is bought. The same reasoning shows that you wouldn't pay 3 cents for it since the person to whom you would have to sell it for 2 cents would object to buying it at that price since he wouldn't be able to sell it for 1 cent. Likewise for prices of 4 cents, 5 cents, 6 cents, and so on. We can use mathematical induction to formalize this argument, which proves conclusively that you wouldn't buy the genie in the bottle for any amount of money. Yet you would almost certainly buy it for $1,000. I know I would. At what point does the argument against buying the bottle cease to be compelling? (I'm ignoring the possibility of foreign currencies that have coins worth less than a penny. This is an American genie.)

The question is more than academic since in countless situations people prepare exclusively for near-term outcomes and don't look very far ahead. They myopically discount the future at an absurdly steep rate.

Average Riches, Likely Poverty

Combining time and money can yield unexpected results in a rather different way. Think back again to the incandescent stock market of the late 1990s and the envious feeling many had that everyone else was making money. You might easily have developed that impression from reading about investing in those halcyon days. In every magazine or newspaper you picked up, you were apt to read about IPOs, the initial public offerings of new companies, and the investment gurus who claimed that they could make your $10,000 grow to more than a million in a year's time. (All right, I'm exaggerating their exaggerations.) But in those same periodicals, even then, you also would have read stories about new companies that were stillborn and naysayers' claims that most investors would lose their $10,000 as well as their shirts by investing in such volatile offerings.

Here's a scenario that helps to illuminate and reconcile such seemingly contradictory claims. Hang on for the math that follows. It may be a bit counterintuitive, but it's not difficult to follow and it illustrates the crucial difference between the arithmetic mean and the geometric mean of a set of returns. (For the record: The arithmetic mean of N different rates of return is what we normally think of as their average; that is, their sum divided by N. The geometric mean of N different rates of return is equal to that rate of return that, if received N times in succession, would be equivalent to receiving the N different rates of return in succession. We can use the formula for compound interest to derive the technical definition. Doing so, we would find that the geometric mean is equal to the Nth root of the product [(1 + first return) × (1 + second return) × (1 + third return) × . . . (1 + Nth return)] − 1.)

Hundreds of IPOs used to come out each year. (Pity that this is only an illustrative flashback.) Let's assume that the first week after the stock comes out, its price is usually extremely volatile. It's impossible to predict which way the price will move, but we'll assume that for half of the companies' offerings the price will rise 80 percent during the first week and for half of the offerings the price will fall 60 percent during this period.

The investing scheme is simple: Buy an IPO each Monday morning and sell it the following Friday afternoon. About half the time you'll earn 80 percent in a week and half the time you'll lose 60 percent in a week for an average gain of 10 percent per week: [(80%) + (−60%)]/2, the arithmetic mean.

Ten percent a week is an amazing average gain, and it's not difficult to determine that after a year of following this strategy, the average worth of an initial $10,000 investment is more than $1.4 million! (Calculation below.) Imagine the newspaper profiles of happy day traders, or week traders in this case, who sold their old cars and turned the proceeds into almost a million and a half dollars in a year.

But what is the most likely outcome if you were to adopt this scheme and the assumptions above held? The answer is that your $10,000 would likely be worth all of $1.95 at the end of a year! Half of all investors adopting such a scheme would have less than $1.95 remaining of their $10,000 nest egg. This same $1.95 is the result of your money growing at a rate equal to the geometric mean of 80 percent and –60 percent over the 52 weeks. (In this case that's equal to the square root (the Nth root for N = 2) of the product [(1 + 80%) × (1 + (–60%))] minus 1, which is the square root of [1.8 × .4] minus 1, which is .85 minus 1, or –.15, a loss of approximately 15 percent each week.)

Before walking through this calculation, let's ask for the intuitive reason for the huge disparity between $1.4 million and $1.95. The answer is that the typical investor will see his investment rise by 80 percent for approximately 26 weeks and decline by 60 percent for 26 weeks. As shown below, it's not difficult to calculate that this results in $1.95 of your money remaining after one year.

The lucky investor, by contrast, will see his investment rise by 80 percent for considerably more than 26 weeks. This will result in astronomical returns that pull the average up. The investments of the unlucky investors will decline by 60 percent for considerably more than 26 weeks, but their losses cannot exceed the original $10,000.

In other words, the enormous returns associated with disproportionately many weeks of 80 percent growth skew the average way up, while even many weeks of 60 percent shrinkage can't drive an investment's value below $0.

In this scenario the stock gurus and the naysayers are both right. The average worth of your $10,000 investment after one year is $1.4 million, but its most likely worth is $1.95.

Which results are the media likely to focus on?

The following example may help clarify matters. Let's examine what happens to the $10,000 in the first two weeks. There are four equally likely possibilities. The investment can increase both weeks, increase the first week and decrease the second, decrease the first week and increase the second, or decrease both weeks. (As we saw in the section on interest theory, an increase of 80 percent is equivalent to multiplying by 1.8. A 60 percent fall is equivalent to multiplying by 0.4.) One-quarter of investors will see their investment increase by a factor of 1.8 × 1.8, or 3.24. Having increased by 80 percent two weeks in a row, their $10,000 will be worth $10,000 × 1.8 × 1.8, or $32,400 in two weeks. One-quarter of investors will see their investment rise by 80 percent the first week and decline by 60 percent the second week. Their investment changes by a factor of 1.8 × 0.4, or 0.72, and will be worth $7,200 after two weeks. Similarly, $7,200 will be the outcome for one-quarter of investors who will see their investment decline the first week and rise the second week, since 0.4 × 1.8 is the same as 1.8 × 0.4. Finally, the unlucky one-quarter of investors whose investment loses 60 percent of its worth for two weeks in a row will have 0.4 × 0.4 × $10,000, or $1,600 after two weeks.

Adding $32,400, $7,200, $7,200, and $1,600 and dividing by 4, we get $12,100 as the average worth of the investments after the first two weeks. That's an average return of 10 percent weekly, since $10,000 × 1.1 × 1.1 = $12,100. More generally, the stock rises an average of 10 percent every week (the average of an 80 percent gain and a 60 percent loss, remember). Thus after 52 weeks, the average value of the investment is $10,000 × $(1.10)^{52}$, which is $1,420,000.

The most *likely* result is that the companies' stock offerings will rise during 26 weeks and fall during 26 weeks. This means that the most likely worth of the investment is $10,000 \times (1.8)^{26} \times (.4)^{26}$, which is only \$1.95. And the geometric mean of 80 percent and –60 percent? Once again, it is the square root of the product of $[(1 + .8) \times (1 - .6)]$ minus 1, which equals approximately –.15. Every week, on average, your portfolio loses 15 percent of its value, and $10,000 \times (1 - .15)^{52}$ equals approximately \$1.95.

Of course, by varying these percentages and time frames, we can get different results, but the principle holds true: The arithmetic mean of the returns far outstrips the geometric mean of the returns, which is also the median (middle) return as well as the most common return. Another example: If half of the time your investment doubles in a week, and half of the time it loses half its value in a week, the most likely outcome is that you'll break even. But the arithmetic mean of your returns is 25 percent per week—$[100\% + (-50\%)]/2$, which means that your initial stake will be worth $10,000 \times 1.25^{52}$, or more than a billion dollars! The geometric mean of your returns is the square root of $(1 + 1) \times (1 - .5)$ minus 1, which is a 0 percent rate of return, indicating that you'll probably end up with the \$10,000 with which you began.

Although these are extreme and unrealistic rates of return, these examples have much more general importance than it might appear. They explain why a majority of investors receive worse-than-average returns and why some mutual fund companies misleadingly stress their average returns. Once again, the reason is that the average or arithmetic mean of different rates of return is always greater than the geometric mean of these rates of return, which is also the median rate of return.

Fat Stocks, Fat People, and P/E

You get what you pay for. As noted, fundamentalists believe that this maxim extends to stock valuation. They argue that a company's stock is worth only what it returns to its holder in dividends and price increases. To determine what that value is, they try to make reasonable estimates of the amount of cash the stock will generate over its lifetime, and then they discount this stream of payments to the present. And how do they estimate these dividends and stock price increases? Value investors tend to use the company's stream of earnings as a reasonable substitute for the stream of dividends paid to them since, the reasoning goes, the earnings are, or eventually will be, paid out in dividends. In the meantime, earnings may be used to grow the company or retire debt, which also increases the company's value. If the earnings of the company are good and promise to get better, and if the economy is growing and interest rates stay low, then high earnings justify paying a lot for a stock. And if not, not.

Thus we have a shortcut for determining a reasonable price for a stock that avoids complicated estimations and calculations: the stock's so-called P/E ratio. You can't look at the business section of a newspaper or watch a business show on TV without hearing constant references to it. The ratio is just that—a ratio or fraction. It's determined by dividing the price P of a share of the company's stock by the company's earnings per share E (usually over the past year). Stock analysts discuss countless ratios, but the P/E ratio, sometimes called simply the multiple, is the most common.

The share price, P, is discovered simply by looking in a newspaper or online, and the earnings per share, E, is obtained by taking the company's total earnings over the past year and dividing it by the number of shares outstanding. (Unfortunately, earnings are not nearly as cut-and-dried as many once thought. All sorts of dodges, equivocations, and outright lies make it a rather plastic notion.)

So how does one use this information? One very common way to interpret the P/E ratio is as a measure of investors' expectations of future earnings. A high P/E indicates high expectations about the company's future earnings, and a low one low expectations. A second way to think of the ratio is simply as the price you must pay to receive (indirectly via dividends and price appreciation) the company's earnings. The P/E ratio is thus both a sort of prediction and an appraisal of the company.

A company with a high P/E must perform to maintain its high ratio. If its earnings don't continue to grow, its price will decline. Consider Microsoft, whose P/E was somewhere north of 100 a few years ago. Today its P/E is under 50, although it's one of the larger companies in Redmond, Washington. Still a goliath, it's nevertheless growing more slowly than it did in its early days. This shrinking of the P/E ratio occurs naturally as start-ups become blue-chip pillars of the business community.

(The pattern of change in a company's growth rate brings to mind a mathematical curve—the S-shaped or logistic curve. This curve seems to characterize a wide variety of phenomena, including the demand for new items of all sorts. Its shape can most easily be explained by imagining a few bacteria in a petri dish. At first the number of bacteria will increase slowly, then at a more rapid exponential rate because of the rich nutrient broth and the ample space in which to expand. Gradually, however, as the bacteria crowd each other, their rate of increase slows and their number stabilizes, at least until the dish is enlarged.

The curve appears to describe the growth of entities as disparate as a composer's symphony production, the rise of airline traffic, highway construction, mainframe computer installations, television ownership, even the building of Gothic cathedrals. Some have speculated that there is a kind of universal principle governing many natural and human phenomena, including the growth of successful businesses.)

Of course, the P/E ratio by itself does not prove anything. A high P/E does not necessarily indicate that a stock is overvalued (too expensive for the cash flow it's likely to generate) and a candidate for selling, nor does a low one indicate that a stock is undervalued and a candidate for buying. A low P/E might mean that a company is in financial hot water despite its earnings.

As WorldCom approached bankruptcy, for example, it had an extremely low P/E ratio. A constant stream of postings in the chatrooms compared it to the P/Es of SBC, AT&T, Deutsche Telekom, Bell South, Verizon, and other comparable companies, which were considerably higher. The stridency of the postings increased when they failed to have their desired effect: Investors hitting their foreheads with the sudden realization that WCOM was a great buy. The posters did have a point, however. One should compare a company's P/E to its value in the past, to that of similar companies, and to the ratios for the sector and the market as a whole. The average P/E for the entire market ranges somewhere between 15 and 25, although there are difficulties with computing such an average. Companies that are losing money, for example, have negative P/Es although they're generally not reported as such; they probably should be. Despite the recent market sell-offs in 2001–2002, some analysts believe that stocks are still too expensive for the cash flow they're likely to generate.

Like other tools that fundamental analysts employ, the P/E ratio seems to be precise, objective, and quasi-mathematical. But, as noted, it too is subject to events in the economy as a whole, strong economies generally supporting higher P/Es. As bears reiteration (verb appropriate), the P in the numerator is not invulnerable to psychological factors nor is the E in the denominator invulnerable to accountants' creativity.

The P/E ratio does provide a better measure of a company's financial health than does stock price alone, just as, for example, the BMI or body mass index (equal to your weight divided by the square of your height in appropriate units) gives a better measure of somatic health than does weight alone. The BMI also suggests other ratios, such as the P/E^2 or, in general, the P/E^X, whose study might exercise analysts to such a degree that their BMIs would fall.

(The parallel between diet and investment regimens is not that far-fetched. There are a bewildering variety of diets and market strategies, and with discipline you can lose weight or make money on most of them. You can diet or invest on your own or pay a counselor who charges a fee and offers no guarantee. Whether the diet or strategy is optimal or not is another matter, as is whether the theory behind the diet or strategy makes sense. Does the diet result in faster, more easily sustained weight loss than the conventional counsel of more exercise and a smaller but balanced intake? Does the market strategy make any excess returns, over and above what you would earn with a blind index fund? Unfortunately, most Americans' waistlines in recent years have been expanding, while their portfolios have been getting slimmer.

Numerical comparisons of the American economy to the world economy are common, but comparisons of our collective weight to that of others are usually just anecdotal. Although we constitute a bit under 5 percent of the world's population, we make up, I suspect, a significantly greater percentage of the world's human biomass.)

There is one refinement of the P/E ratio that some find very helpful. It's called the PEG ratio and it is the P/E ratio divided by (100 times) the expected annual growth rate of earnings. A low PEG is usually taken to mean that the stock is undervalued, since the growth rate of earnings is high relative to the P/E. High P/E ratios are fine if the rate of growth of the company is sufficiently rapid. A high-tech company with a P/E ratio of 80 and annual growth of 40 percent will have a PEG of

2 and may sound promising, but a stodgier manufacturing company with a P/E of 7 and an earnings growth rate of 14 percent will have a more attractive PEG of .5. (Once again, negative values are excluded.)

Some investors, including the Motley Fool and Peter Lynch, recommend buying stocks with a PEG of .5 or lower and selling stocks with PEG of 1.50 or higher, although with a number of exceptions. Of course, finding stocks having such a low PEG is no easy task.

Contrarian Investing and
the *Sports Illustrated* Cover Jinx

As with technical analysis, the question arises: Does it work? Does using the ideas of fundamental analysis enable you to do better than you would by investing in a broad-gauged index fund? Do stocks deemed undervalued by value investors constitute an exception to the efficiency of the markets? (Note that the term "undervalued" itself contests the efficient market hypothesis, which maintains that all stocks are always valued just right.)

The evidence in favor of fundamental analysis is a bit more compelling than that supporting technical analysis. Value investing does seem to yield moderately better rates of return. A number of studies have suggested, for example, that stocks with low P/E ratios (undervalued, that is) yield better returns than do those with high P/E ratios, the effect's strength varying with the type and size of the company. The notion of risk, discussed in chapter 6, complicates the issue.

Value investing is frequently contrasted with growth investing, the chasing of fast-growing companies with high P/Es. It brings better returns, according to some of its supporters, because it benefits from investors' overreactions. Investors sign on too quickly to the hype surrounding fast-growing companies and underestimate the prospects of solid, if humdrum companies of the type that Warren Buffett likes—Coca-Cola, for instance. (I write this in a study littered with empty cans of Diet Coke.)

The appeal of value investing tends to be contrarian, and many of the strategies derived from fundamental analysis reflect this. The "dogs of the Dow" strategy counsels investors to buy the ten Dow stocks (among the thirty stocks that go into the Dow-Jones Industrial Average) whose price-to-dividend, P/D, ratios are the lowest. Dividends are not earnings, but the strategy corresponds very loosely to buying the ten stocks with the lowest P/E ratios. Since the companies are established organizations, the thinking goes, they're unlikely to go bankrupt and thus their relatively poor performance probably indicates that they're temporarily undervalued. This strategy, again similar to one promoted by the Motley Fool, became popular in the late '80s and early '90s and did result in greater gains than those achieved by, say, the broad-gauged S&P 500 average. As with all such strategies, however, the increased returns tended to shrink as more people adopted it.

A ratio that seems to be more strongly related to increased returns than price-to-dividends or price-to-earnings is the price-to-book ratio, P/B. The denominator B is the company's book value per share—its total assets minus the sum of total liabilities and intangible assets, divided by the number of shares. The P/B ratio changes less over time than does the P/E ratio and has the further virtue of almost always being positive. Book value is meant to capture something basic about a company, but like earnings it can be a rather malleable number.

Nevertheless, a well-known and influential study by the economists Eugene Fama and Ken French has shown P/B to be a useful diagnostic device. The authors focused on the period from 1963 to 1990 and divided almost all the stocks on the

New York Stock Exchange and the Nasdaq into ten groups: the 10 percent of the companies with the highest P/B ratios, the 10 percent with the next highest, on down to the 10 percent with the lowest P/B ratios. (These divisions are called deciles.) Once again a contrarian strategy achieved better than average rates of return. Without exception, every decile with lower P/B ratios outperformed the deciles with higher P/B ratios. The decile with lowest P/B ratios had an average return of 21.4 percent versus 8 percent for the decile with the highest P/B ratios. Other studies' findings have been similar, although less pronounced. Some economists, notably James O'Shaughnessy, claim that a low price to sales ratio, P/S, is an even stronger predictor of better-than-average returns.

Concern with the fundamental ratios of a company is not new. Finance icons Benjamin Graham and David Dodd, in their canonical 1934 text *Security Analysis,* stressed the importance of low P/E and P/B ratios in selecting stocks to buy. Some even stipulate that low ratios constitute the definition of "value stocks" and that high ratios define "growth stocks." There are more nuanced definitions, but there is a consensus that value stocks typically include most of those in oil, finance, utilities, and manufacturing, while growth stocks typically include most of those in computers, telecommunications, pharmaceuticals, and high technology.

Foreign markets seem to deliver value investors the same excessive returns. Studies that divide a country's stocks into fifths according to the value of their P/E and P/B ratios, for example, have generally found that companies with low ratios had higher returns than those with high ratios. Once again, over the next few years, the undervalued, unpopular stocks performed better.

There are other sorts of contrarian anomalies. Richard Thaler and Werner DeBondt examined the thirty-five stocks on the New York Stock Exchange with the highest rates of returns and the thirty-five with the lowest rates for each year from the 1930s until the 1970s. Three to five years later, the best performers had average returns lower than those of the NYSE, while the worst performers had averages considerably higher than the index. Andrew Lo and Craig MacKinlay, as mentioned earlier, came to similar contrarian conclusions more recently, but theirs were significantly weaker, reflecting perhaps the increasing popularity and hence decreasing effectiveness of contrarian strategies.

Another result with a contrarian feel derives from management guru Tom Peters's book *In Search of Excellence,* in which he deemed a number of companies "excellent" based on various fundamental measures and ratios. Using these same measures a few years after Peters's book, Michelle Clayman compiled a list of "execrable" companies (my word, not hers) and compared the fates of the two groups of companies. Once again there was a regression to the mean, with the execrable companies doing considerably better than the excellent ones five years after being so designated.

All these contrarian findings underline the psychological importance of a phenomenon I've only briefly mentioned: regression to the mean. Is the decline of Peters's excellent companies, or of other companies with good P/E and P/B ratios the business analogue of the *Sports Illustrated* cover jinx?

For those who don't follow sports (a field of endeavor where the numbers are usually more trustworthy than in business), a black cat stared out from the cover of the January 2002 issue of *Sports Illustrated* signaling that the lead article was about the magazine's infamous cover jinx. Many fans swear that getting on the cover of the magazine is a prelude to a fall from grace, and much of the article detailed instances of an athlete's or a team's sudden decline after appearing on the cover.

There were reports that St. Louis Rams quarterback Kurt Warner turned down an offer to pose with the black cat on the issue's cover. He wears No. 13 on his back, so maybe there's a limit to how much bad luck he can withstand. Besides, a couple of weeks after gracing the cover in October 2000, Warner broke his little finger and was sidelined for five games.

The sheer number of cases of less than stellar performance or worse following a cover appearance is impressive at first. The author of the jinx story, Alexander Wolff, directed a team of researchers who examined almost all of the magazine's nearly 2,500 covers dating back to the first one, featuring Milwaukee Braves third baseman Eddie Mathews in August 1954. Mathews was injured shortly after that. In October 1982, Penn State was unbeaten and the cover featured its quarterback, Todd Blackledge. The next week Blackledge threw four interceptions against Alabama and Penn State lost big. The jinx struck Barry Bonds in late May 1993, seeming to knock him into a dry spell that reduced his batting average forty points in just two weeks.

I'll stop. The article cited case after case. More generally, the researchers found that within two weeks of a cover appearance, over a third of the honorees suffered injuries, slumps, or other misfortunes. Theories abound on the cause of the cover jinx, many having to do with players or teams choking under the added performance pressure.

A much better explanation is that no explanation is needed. It's what you would expect. People often attribute meaning to phenomena governed only by a regression to the mean, the mathematical tendency for an extreme value of an at least partially chance-dependent quantity to be followed by a value closer to the average. Sports and business are certainly chancy enterprises and thus subject to regression. So is genetics to an extent, and so very tall parents can be expected to have offspring who are tall, but probably not as tall as they are. A similar tendency holds for the children of very short parents.

If I were a professional darts player and threw one hundred darts at a target (or a list of companies in a newspaper's business section) during a tournament and managed to hit the bull's-eye (or a rising stock) a record-breaking eighty-three times, the next time I threw one hundred darts, I probably wouldn't do nearly as well. If featured on a magazine cover (*Sports Illustrated* or *Barron's*) for the eighty-three hits, I'd probably be adjudged a casualty of the jinx too.

Regression to the mean is widespread. The sequel to a great CD is usually not as good as the original. The same can be said of the novel after the best-seller, the proverbial sophomore slump, Tom Peters's excellent companies faring relatively badly after a few good years, and, perhaps, the fates of Bernie Ebbers of WorldCom, John Rigas of Adelphia, Ken Lay of Enron, Gary Winnick of Global Crossing, Jean-Marie Messier of Vivendi (to throw in a European), Joseph Nacchio of Qwest, and Dennis Kozlowski of Tyco—all CEOs of large companies who received adulatory coverage before their recent plunges from grace. (Satirewire.com refers to these publicity-fleeing, company-draining executives as the CEOnistas.)

There is a more optimistic side to regression. I suggest that *Sports Illustrated* consider featuring an established player who has had a particularly bad couple of months on its *back* cover. Then they could run feature stories on the *boost* associated with such appearances. *Barron's* could do the same thing with its back cover.

An expectation of a regression to the mean is not the whole story, of course, but there are dozens of studies suggesting that value investing, generally over a three-to-five year period, does result in better rates of return than, say, growth investing. It's important to remember, however, that the size of the effect varies with the study (not surprisingly, some studies find zero or a negative effect), transaction costs can eat up some or all of it, and competing investors tend to shrink it over time.

In chapter 6 I'll consider the notion of risk in general, but there is a particular sort of risk that may be relevant to value stocks. Invoking the truism that higher risks bring greater returns even in an efficient market, some have argued that value companies are risky because they're so colorless and easily ignored that their stock prices must be lower to compensate! Using "risky" in this way is risky, however, since it seems to explain too much and hence nothing at all.

Accounting Practices, WorldCom's Problems

Even if value investing made better sense than investing in broad-gauged index funds (and that is certainly not proved) a big problem remains. Many investors lack a clear understanding of the narrow meanings of the denominators in the P/E, P/B, and P/D ratios, and an uncritical use of these ratios can be costly.

People are easily bamboozled about numbers and money even in everyday circumstances. Consider the well-known story of the three men attending a convention at a hotel. They rent a booth for $30, and after they go to their booth, the manager realizes that it costs only $25 and that he's overcharged them. He gives $5 to the bellhop and directs him to give it back to the three men. Not knowing how to divide the $5 evenly, the bellhop decides to give $1 to each of the three men and pockets the remaining $2 for himself. Later that night the bellhop realizes that the men each paid $9 ($10 minus the $1 they received from him). Thus, since the $27 the men paid (3 × $9 = $27) plus the $2 that he took for himself sums to $29, the bellhop wonders what happened to the missing dollar. What did happen to it?

The answer, of course, is that there is no missing dollar. You can see this more easily if we assume that the manager originally made a bigger mistake, realizing after charging the men $30 that the booth costs only $20 and that he's overcharged them $10. He gives $10 to the bellhop and directs him to give it back to the three men. Not knowing how to divide the $10 evenly, the bellhop decides to give $3 to each of the three men and pockets the remaining $1 for himself. Later that night the bellhop realizes that the men each paid $7 ($10 minus the $3 they received from him). Thus, since the $21 the men paid (3 × $7 = $21) plus the $1 he took for himself sums to $22, the bellhop wonders what happened to the missing $8. In this case there's less temptation to think that there's any reason the sum should be $30.

If people are baffled by these "disappearances," and many are, what makes us so confident that they understand the accounting intricacies on the basis of which they may be planning to invest their hard-earned (or even easily earned) dollars? As the recent accounting scandals make clear, even a good understanding of these notions is sometimes of little help in deciphering the condition of a company's finances. Making sense of accounting documents and seeing how balance sheets, cash flow statements, and income statements feed into each other is not something investors often do. They rely instead on analysts and auditors, and this is why conflating the latter roles with those of investment bankers and consultants causes such concern.

If an accounting firm auditing a company also serves as a consultant to the company, there is a troubling conflict of interest. (A similar crossing of professional lines that is more upsetting to me has been curtailed by Eliot Spitzer, New York attorney general. One typical instance involved Jack Grubman, arguably the most influential analyst of telecommunications companies such as WorldCom, who was incestually entangled in the investments and underwriting of the very companies he was supposed to be dispassionately analyzing.)

A student's personal tutor who is paid to improve his or her performance should not also be responsible for grading the student's exams. Nor should an athlete's personal coach be the referee in a game in which the athlete competes. The situation may not be exactly the same since, as accounting firms have argued, different departments are involved in auditing and consulting. Nevertheless, there is at least the appearance of impropriety, and often enough the reality too.

Such improprieties come in many flavors. Enron's accounting feints and misdirections involving off-shore entities and complicated derivatives trading were at least subtle and almost elegant. WorldCom's moves, by contrast, were so simple and blunt that Arthur Andersen's seeming blindness is jaw-dropping. Somehow Andersen's auditors failed to note that WorldCom had classified $3.8 billion in corporate expenses as capital investments. Since expenses are charged against profits as they are incurred, while capital investments are spread out over many years, this accounting "mistake" allowed WorldCom to report profits instead of losses for at least two years and probably longer. After this revelation, investigators learned that earnings were increased another $3.3 billion by some combination of the same ruse and the shifting of funds from exaggerated one-time charges against earnings (bad debts and the like) back into earnings as the need arose, creating, in effect, a huge slush fund. Finally (almost finally?) in November 2002 the SEC charged WorldCom with inflating earnings by an additional $2 billion, bringing the total financial misstatements to over $9 billion! (Many comparisons with this sum are possible; one is that $9 billion is more than twice the gross domestic product of Somalia.)

WorldCom's accounting fraud first came to light in June 2002, long after I had invested a lot of money in the company and passively watched as its value shriveled to almost nothing. Bernie Ebbers and company had not merely made $1 disappear as in the puzzle above, but had presided over the vanishing of approximately $190 billion, the value of WorldCom's market capitalization in 1999—$64 a share times 3 billion shares. For this and many other reasons it might be argued that both the multi-trillion-dollar boom of the '90s and the comparably sized bust of the early '00s were largely driven by telecommunications. (With such gargantuan numbers it's important to remember the fundamental laws of financial estimation: A trillion dollars plus or minus a few dozen billion is still a trillion, just as a billion dollars plus or minus a few dozen million is still a billion.)

I was a victim, but the primary victimizer, I'm sorry to say, was not WorldCom management but myself. Putting so much money into one stock, failing to place stop-loss orders or to buy insurance puts, and investing on margin (puts and margin will be discussed in chapter 6) were foolhardy and certainly not based by the company's fundamentals. Besides, these fundamentals and other warning signs should have been visible even through the accounting smoke screen.

The primary indication of trouble was the developing glut in the telecommunications industry. Several commentators have observed that the industry's trajectory over the last decade resembled that of the railroad industry after the Civil War. The opening of the West, governmental inducements, and new technology led the railroads to build thousands of miles of unneeded track. They borrowed heavily, each company attempting to be the dominant player; their revenue couldn't keep pace with the rising debt; and the resulting collapse brought on an economic depression in 1873.

Substitute fiber-optic cable for railroad tracks, the opening of global markets for the opening of the West, the Internet for the intercontinental railroad network, and governmental inducements for governmental inducements, and there you have it. Millions of miles of unused fiber-optic cable costing billions of dollars were laid to capture the insufficiently burgeoning demand for online music and pet stores. In a nutshell: Debts increased, competition grew keener, revenue declined, and bankruptcies loomed. Happily, however, no depression, at least as of this writing.

In retrospect, it's clear that the situation was untenable and that WorldCom's accounting tricks and deceptions (as well as Global Crossing's and others') merely papered over what would soon have come to light anyway: These companies were losing a lot of money. Still, anyone can be forgiven for not recognizing the problem of overcapacity or for not seeing through the hype and fraudulent accounting. (Far less blameless, if I may self-flagellate again, were my dumb investing practices, for which WorldCom management and accountants certainly weren't responsible.) The real source of most people's dismay and apprehension, I suspect, derives less from accountants' malfeasance than from the market's continuing to flounder. If it were rising, interest in the various accounting reforms that have been proposed and enacted would rival the public's keen fascination with partial differential equations or Cantor's continuum hypothesis.

Reforms can only accomplish so much. There are countless ways for accountants to dissemble, many of which shade into legitimate moves, and this highlights a different tension running through the accounting profession. The precision and objectivity of its bookkeeping fit uncomfortably alongside the vagueness and subjectivity of many of its practices. Every day accountants must make judgments and determinations that are debatable—about the way to value inventory, the burdens of pensions and health care, the quantification of goodwill, the cost of warranties, or the classification of expenses—but once made, these judgments result in numbers, exact to the nearest penny, that seem indubitable.

The situation is analogous to that in applied mathematics where the appropriateness of a mathematical model is always vulnerable to criticism. Is this model the right one for this situation? Are these assumptions warranted? Once the assumptions are made and the model is adopted, however, the numbers and organizational clarity that result have an irresistible appeal. Responding to this appeal two hundred years ago, the German poet Goethe rapturously described accounting this way: "Double entry bookkeeping is one of the most beautiful discoveries of the human spirit."

Focusing only on the bookkeeping and the numerical output, however, and refusing to examine the legitimacy of the assumptions made, can be disastrous, both in mathematics and accounting. Recall the tribe of bear hunters who became extinct once they became expert in the complex calculations of vector analysis. Before they encountered mathematics, the tribesmen killed, with their bows and arrows, all the bears they could eat. After mastering vector analysis, they starved. Whenever they spotted a bear to the northeast, for example, they would fire, as vector analysis suggested, one arrow to the north and one to the east.

Even more important than the appropriateness of accounting rules and models is the transparency of these practices. It makes compelling sense, for example, for companies to count the stock options given to executives and employees as expenses. Very few do so, but as long as everyone knows this, the damage is not as great as it could be. Everyone knows what's going on and can adapt to it.

If an accounting practice is transparent, then an outside auditor who is independent and trusted can, when necessary, issue a statement analogous to the warning made by the independent and trusted matriarch from chapter 1. By making a bit of information common knowledge, an auditor (or the SEC) can alert everyone involved to a violation and stimulate remedial action. If the auditor is not independent or not trusted (as Harvey Pitt, the recently departed chairman of the SEC, was not), then he is simply another player and violations, although perhaps widely and mutually known, will not become commonly known (everyone knowing that everyone else knows it and knowing that everyone else knows they know it and so on), and no action will result. In a similar way, family secrets take on a different character and have some hope of being resolved when they become common knowledge rather than merely mutual knowledge. Family and corporate "secrets" (such as WorldCom's misclassification of expenses) are often widely known, just not talked about.

Transparency, trust, independence, and authority are all needed to make the accounting system work. They are all in great demand, but sometimes in short supply.

6 | Options, Risk, and Volatility

C onsider a rather ugly mathematical physicist who goes to the same bar every evening, always takes the second to the last seat, and seems to speak toward the empty seat next to his as if someone were there. The bartender notes this, and on Valentine's Day when the physicist seems to be especially fervent in his conversation, he asks why he is talking into the air. The physicist scoffs that the bartender doesn't know anything about quantum mechanics. "There is no such thing as a vacuum. Virtual particles flit in and out of existence, and there is a non-zero probability that a beautiful woman will materialize and, when she does, I want to be here to ask her out." The bartender is baffled and asks why the physicist doesn't just ask one of the real women who are in the bar. "You never know. One of them might say yes." The physicist sneers, "Do you know how unlikely that is?"

Being able to estimate probabilities, especially minuscule ones, is essential when dealing with stock options. I'll soon describe the language of puts and calls, and we'll see why the January 2003 calls on WCOM at 15 have as much chance of ending up in the money as Britney Spears has of suddenly materializing before the ugly physicist.

Options and the Calls of the Wild

Here's a thought experiment: Two people (or the same person in parallel universes) have roughly similar lives until each undertakes some significant endeavor. The endeavors are equally worthy and equally likely to result in success, but one endeavor ultimately leads to good things for X and his family and friends, and the other leads to bad things for Y and his family and friends. It seems that X and Y should receive roughly comparable evaluations for their decision, but generally they won't. Unwarranted though it may be, X will be judged kindly and Y harshly. I tell this in part because I'd like to exonerate myself for my investing behavior by claiming status as a faultless Mr. Y, but I don't qualify.

By late January 2002, WCOM had sunk to about $10 per share, and I was feeling not only dispirited but guilty about losing so much money on it. Losing money in the stock market often induces guilt in those who have lost it, whether they've done anything culpable or not. Whatever your views on the randomness of the market, it's indisputable that chance plays a huge role, so it makes no sense to feel guilty about having called heads when a tails comes up. If this was what I'd done, I could claim to be a Mr. Y: It wouldn't have been my fault. Alas, as I mentioned, it does make sense to blame yourself for betting recklessly on a particular stock (or on options for it).

There is a term used on Wall Street to describe traders and others who "blow up" (that is, lose a fortune) and as a result become hollow, sepulchral figures. The term is "ghost" and I have developed more empathy for ghosts than I wanted to have. Often they achieve their funereal status by taking *unnecessary* risks, risks that they could and should have "diversified away." One perhaps counterintuitive way in which to reduce risk is to buy and sell stock options.

Many people think of stock options as slot machines, roulette wheels, or dark horse long shots; that is, as pure gambles. Others think of them as absurdly large inducements for people to stay with a company or as rewards for taking a company public. I have no argument with these characterizations, but much of the time an option is more akin to a boring old insurance policy. Just as one buys an insurance policy in case one's washing machine breaks down, one often buys options in case one's stock breaks down. They lessen risk, which is the bete noire, bugbear, and bane of investors' lives and the topic of this chapter.

How options work is best explained with a few numerical examples. (How they're misused is reserved for the next section.) Assume that you have 1,000 shares of AOL (just to give WCOM a rest), and it is selling at $20 per share. Although you think it's likely to rise in the long term, you realize there's a chance that it may fall significantly in the next six months. You could insure against this by buying 1,000 "put" options at an appropriate price. These would give you the right to sell 1,000 shares of AOL for, say, $17.50 for the next six months. If the stock rises or falls less than $2.50, the puts become worthless in six months (just as your washing machine warranty becomes worthless on its expiration if your machine has not broken down by then). Your right to sell shares at $17.50 is not attractive if the price of the stock is more than that. However, if the stock plunges to, say, $10 per share within the six-month period, your right to sell shares at $17.50 is worth at least $7.50 per share. Buying put options is a hedge against a precipitous decline in the price of the underlying stock.

As I was first writing this, only a few paragraphs and a few days after WCOM had fallen to $10, it fell to under $8 per share, and I wished I had bought a boatload of puts on it months before when they were dirt cheap.

In addition to put options, there are "call" options. Buying them gives you the right to buy a stock at a certain price within a specified period of time. You might be tempted to buy calls when you strongly believe that a stock, say Intel this time (abbreviated INTC), selling at $25 per share, will rise substantially during the next year. Maybe you can't afford to buy many shares of INTC, but you can afford to buy calls giving you the right to buy shares at, say, $30 during the next year. If the stock falls or rises less than $5 during the next year, the calls become worthless. Your right to buy shares at $30 is not attractive if the price of the stock is less than that. But if the stock rises to, say, $40 per share within the year, each call is worth at least $10. Buying call options is a bet on a substantial rise in the price of the stock. It is also a way to insure that you are not left out when a stock, too expensive to buy outright, begins to take off. (The figures $17.50 and $30 in the AOL and INTC examples above are the "strike" prices of the respective options; this is the price of the stock that determines the point at which the option has intrinsic value or is "in the money.")

One of the most alluring aspects of buying puts and calls is that your losses are limited to what you have paid for them, but the potential gains are unlimited in the case of calls and very substantial in the case of puts. Because of these huge potential gains, options probably induce a comparably huge amount of fantasy—countless investors thinking something like "the option for INTC with a $30 strike price costs around a dollar, so if the stock goes to $45 in the next year, I'll make 15 times my investment. And if it goes to $65, I'll make 35 times my investment." The attraction for some speculators is not much different from that of a lottery.

Although I've often quoted approvingly Voltaire's quip that lotteries are a tax on stupidity (or at least on innumeracy), yes, I did buy a boatload of now valueless WCOM calls. In fact, over the two years of my involvement with the stock, I bought many thousands of January 2003 calls on WCOM at $15. I thought that whatever problems the company had were temporary and that by 2003 it would right itself and, in the process, me as well. Call me an ugly physicist.

There is, of course, a market in puts and calls, which means that people sell them as well as buy them. Not surprisingly, the payoffs are reversed for sellers of options. If you sell calls for INTC with a strike price of $30 that expires in a year, then you keep your proceeds from the sale of the calls and pay nothing unless the stock moves above $30. If, however, the stock moves to, say, $35, you must supply the buyer of the calls with shares of INTC at $30. Selling calls is thus a bet that the stock will either decline or rise only slightly in a given time period. Likewise, selling puts is a bet that the stock will either rise or decline only slightly.

One common investment strategy is to buy shares of a stock and simultaneously sell calls on them. Say, for example, you buy some shares of INTC stock at $25 per share and sell six-month calls on them with a strike price of $30. If the stock price doesn't rise to $30, you keep the proceeds from the sale of the calls, but if the stock price does exceed $30, you can sell your own shares to the buyer of the calls, thus limiting the considerable risk in selling calls. This selling of "covered" calls (covered because you own the stock and don't have to buy it at a high price to satisfy the buyer of the call) is one of many hedges investors can employ to maximize their returns and minimize their risks.

More generally, you can buy and sell the underlying stock and mix and match calls and puts with different expiration dates and strike prices to create a large variety of potential profit and loss outcomes. These combinations go by names like "straddles," "strangles," "condors," and "butterflies," but whatever strange and contorted animal they're named for, like all insurance policies, they cost money. A surprisingly difficult question in finance has been "How does one place a value on a put or a call?" If you're insuring your house, some of the determinants of the policy premium are the replacement cost of the house, the length of time the policy is in effect, and the amount of the deductible. The considerations for a stock include these plus others having to do with the rise and fall of stock prices.

Although the practice and theory of insurance have a long history (Lloyd's of London dates from the late seventeenth century), it wasn't until 1973 that a way was found to rationally assign costs to options. In that year Fischer Black and Myron Scholes published a formula that, although much refined since, is still the basic valuation tool for options of all sorts. Their work and that of Robert Merton won the Nobel prize for economics in 1997.

Louis Bachelier, whom I mentioned in chapter 4, also devised a formula for options more than one hundred years ago. Bachelier's formula was developed in connection with his famous 1900 doctoral dissertation in which he was the first to conceive of the stock market as a chance process in which price movements up and down were normally distributed. His work, which utilized the mathematical theory of Brownian motion, was way ahead of its time and hence was largely ignored. His options formula was also prescient, but ultimately misleading. (One reason for its failure is that Bachelier didn't take account of the effect of compounding on stock returns. Over time this leads to what is called a "lognormal" distribution rather than a normal one.)

The Black-Scholes options formula depends on five parameters: the present price of the stock, the length of time until the option expires, the interest rate, the strike price of the option, and the volatility of the underlying stock. Without getting into the mechanics of the formula, we can see that certain general relations among these parameters are commonsensical. For example, a call that expires two years from now has to cost more than one that expires in three months since the later expiration date gives the stock more time to exceed the strike price. Likewise, a call with a strike price a point or two above the present stock price will cost more than one five points above the stock price. And options on a stock whose volatility is high will cost more than options on stocks that barely move from quarter to quarter (just as a short man on a pogo stick is more likely to be able to peek over a nine-foot fence than a tall man who can't jump). Less intuitive is the fact that the cost of a call option also rises with the interest rate, assuming all other parameters remain unchanged.

Although there are any number of books and websites on the Black-Scholes formula, it and its variants are more likely to be used by professional traders than by gamblers, who rely on commonsense considerations and gut feel. Viewing options as pure bets, gamblers are generally as interested in carefully pricing them as casino-goers are in the payoff ratios of slot machines.

The Lure of Illegal Leverage

Because of the leverage possible with the purchase, sale, or mere possession of options, they sometimes attract people who aren't content to merely play the slots but wish to stick their thumbs onto the spinning disks and directly affect the outcomes. One such group of people are CEOs and other management personnel who stand to reap huge amounts of money if they can somehow contrive (by hook, crook, or, too often, by cooking the books) to raise their companies' stock price. Even if the rise is only temporary, the suddenly valuable call options can "earn" them tens of millions of dollars. This is the luxury version of "pump and dump" that has animated much of the recent corporate malfeasance.

(Such malfeasance might make for an interesting novel. On public television one sometimes sees a fantasia in which diverse historical figures are assembled for an imaginary conversation. Think, for example, of Leonardo da Vinci, Thomas Edison, and Benjamin Franklin discussing innovation. Sometimes a contemporary is added to the mix or simply paired with an illustrious precursor—maybe Karl Popper and David Hume, Stephen Hawking and Isaac Newton, or Henry Kissinger and Machiavelli. Recently I tried to think with whom I might pair a present-day ace CEO, investor, or analyst. There are a number of books about the supposed relevance to contemporary business practices of Plato, Aristotle, and other ancient wise men, but the conversation I'd be most interested in would be one between a current wheeler-dealer and some accomplished hoaxer of the past, maybe Dennis Koslowski and P. T. Barnum, or Kenneth Lay and Harry Houdini, or possibly Bernie Ebbers and Elmer Gantry.)

Option leverage works in the opposite direction as well, the options-fueled version of "short and distort." One particularly abhorrent example may have occurred in connection with the bombing of the World Trade Center. Just after September 11, 2001, there were reports that Al Qaeda operatives in Europe had bought millions of dollars worth of puts on various stock indices earlier in the month, reasoning that the imminent attacks would lead to a precipitous drop in the value of these indices and a consequent enormous rise in the value of their puts. They may have succeeded, although banking secrecy laws in Switzerland and elsewhere make that unclear.

Much more commonly, people buy puts on a stock and then try to depress its price in less indiscriminately murderous ways. A stockbroker friend of mine tells me, for example, of his fantasy of writing a mystery novel in which speculators buy puts on a company whose senior management is absolutely critical to the success of the company. The imaginary speculators then proceed to embarrass, undermine, and ultimately kill the senior management in order to reap the benefit of the soon-to-be valuable puts. The WorldCom chatroom, home to all sorts of utterly baseless rumors, once entertained a brief discussion about the possibility of WorldCom management having been blackmailed into doing all the ill-considered things they did on pain of having some awful secrets revealed. The presumption was that the blackmailers had bought WCOM puts.

Intricacies abound, but the same basic logic governing stock options is at work in the pricing of derivatives. Sharing only the same name as the notion studied in calculus, derivatives are financial instruments whose value is derived from some underlying asset—the stock of a company, commodities like cotton, pork bellies, and natural gas, or almost anything whose value varies significantly over time. They present the same temptation to directly change, affect, or manipulate conditions, and the opportunities for doing so are more varied and would also make for an intriguing business mystery novel.

The leverage involved in trading options and derivatives brings to mind a classic quote from Archimedes, who maintained that given a fulcrum, a long enough lever, and a place to stand, he could move the earth. The world-changing dreams that created the suggestively named WorldCom, Global Crossing, Quantum Group (George Soros' companies, no stranger to speculation), and others may have been similar in scope. The metaphorical baggage of levers and options is telling.

One can also look at seemingly non-financial situations and discern something like the buying, selling, and manipulating of options. For example, the practice of defraying the medical bills of AIDS patients in exchange for being made the beneficiary of their insurance policies has disappeared with the increased longevity of those with AIDS. However, if the deal were modified so that the parties put a time limit on their agreement, it could be considered a standard option sale. The "option buyer" would pay a sum of money, and the patient/option seller would make the buyer the beneficiary for an agreed-upon period of time. If the patient happens not to expire within that time, the "option" does. Maybe another mystery novel here?

Less ghoulish variants of option buying, selling, and manipulating play an important role in everyday life from education and family planning to politics. Political options, better known as campaign contributions to relatively unknown candidates, usually expire worthless after the candidate loses the race. If he or she is elected, however, the "call option" becomes very valuable, enabling the contributor to literally call on the new officeholder. There is no problem with that, but direct manipulation of conditions that might increase the value of the political option is generally called "dirty tricks."

For all the excesses options sometimes inspire, they are generally a good thing, a valuable lubricant that enables prudent hedgers and adventurous gamblers to form a mutually advantageous market. It's only when the option holders do something to directly affect the value of the options that the lure of leverage turns lurid.

Short-Selling, Margin Buying, and Familial Finances

An old Wall Street couplet says, "He who sells what isn't his'n must buy it back or go to prison." The lines allude to "short-selling," the selling of stocks one doesn't own in the hope that the price will decline and one can buy the shares back at a lower price in the future. The practice is very risky because the price might rise precipitously in the interim, but many frown upon short-selling for another reason. They consider it hostile or anti-social to bet that a stock will decline. You can bet that your favorite horse wins by a length, not that some other horse breaks its leg. A simple example, however, suggests that short-selling can be a necessary corrective to the sometimes overly optimistic bias of the market.

Imagine that a group of investors has a variety of attitudes to the stock of company X, ranging from a very bearish 1 through a neutral 5 or 6 to a very bullish 10. In general, who is going to buy the stock? It will generally be those whose evaluations are in the 7 to 10 range. Their average valuation will be, let's assume, 8 or 9. But if those investors in the 1 to 4 range who are quite dubious of the stock were as likely to short sell X as those in the 7 to 10 range were to buy it, then the average valuation might be a more realistic 5 or 6.

Another positive way to look at short-selling is as a way to double the number of stock tips you receive. Tips about a bad stock become as useful as tips about a good one, assuming that you believe any tips. Short-selling is occasionally referred to as "selling on margin," and it is closely related to "buying on margin," the practice of buying stock with money borrowed from your broker.

To illustrate the latter, assume you own 5,000 shares of WCOM and it's selling at $20 per share (ah, remembrance of riches past). Since your investment in WCOM is worth $100,000, you can borrow up to this amount from your broker and, if you're very bullish on WCOM and a bit reckless, you can use it to buy an additional 5,000 shares on margin, making the total market value of your WCOM holdings $200,000 ($20 × 10,000 shares). Federal regulations require that the amount you owe your broker be no more than 50 percent of the total market value of your holdings. (Percentages vary with the broker, stock, and type of account.) This is no problem if the price of WCOM rises to $25 per share, since the $100,000 you owe your broker will then constitute only 40 percent of the $250,000 ($25 × 10,000) market value of your WCOM shares. But consider what happens if the stock falls to $15 per share. The $100,000 you owe now constitutes 67 percent of the $150,000 ($15 × 10,000) market value of your WCOM shares, and you will receive a "margin call" to deposit immediately enough money ($25,000) into your account to bring you back into compliance with the 50 percent requirement. Further declines in the stock price will result in more margin calls.

I'm embarrassed to reiterate that my devotion to WCOM (others may characterize my relationship to the stock in less kindly terms) led me to buy it on margin and to make the margin calls on it as it continued its long, relentless decline. Receiving a margin call (which often takes the literal form of a telephone call) is, I can attest, unnerving and confronts you with a stark choice. Sell your holdings and get out of the game now or quickly scare up some money to stay in it.

My first margin call on WCOM is illustrative. Although the call was rather small, I was leaning toward selling some of my shares rather than depositing yet more money in my account. Unfortunately (in retrospect), I needed a book quickly and decided to go to the Borders store in Center City, Philadelphia, to look for it. While doing so, I came across the phrase "staying in the game" while browsing and realized that staying in the game was what I still wanted to do. I realized too that Schwab was very close to Borders and that I had a check in my pocket.

My wife was with me, and though she knew of my investment in WCOM, at the time she was not aware of its extent nor of the fact that I'd bought on margin. (Readily granting that this doesn't say much for the transparency of my financial practices, which would not likely be approved by even the most lax Familial Securities Commission, I plead guilty to spousal deception.) When she went upstairs, I ducked out of the store and made the margin call. My illicit affair with WCOM continued. Occasionally exciting, it was for the most part anxiety-inducing and pleasureless, not to mention costly.

I took some comfort from the fact that my margin buying distantly mirrored that of WorldCom's Bernie Ebbers, who borrowed approximately $400 million to buy WCOM shares. (More recent allegations have put his borrowings at closer to $1 billion, some of it for personal reasons unrelated to World-Com. Enron's Ken Lay, by contrast, borrowed only $10 to $20 million.) When he couldn't make the ballooning margin calls, the board of directors extended him a very low interest loan that was one factor leading to further investor unrest, massive sell-offs, and more trips to Borders for me.

Relatively few individuals short-sell or buy on margin, but the practice is very common among hedge funds—private, lightly regulated investment portfolios managed by people who employ virtually every financial tool known to man. They can short-sell, buy on margin, use various other sorts of leverage, or engage in complicated arbitrage (the near simultaneous buying and selling of the same stock, bond, commodity, or anything else, in order to profit from tiny price discrepancies). They're called "hedge funds" because many of them try to minimize the risks of wealthy investors. Others fail to hedge their bets at all.

A prime example of the latter is the collapse in 1998 of Long-Term Capital Management, a hedge fund, two of whose founding partners, Robert Merton and Myron Scholes, were the aforementioned Nobel prize winners who, together with Fischer Black, derived the celebrated formula for pricing options. Despite the presence of such seminal thinkers on the board of LTCM, the debacle roiled the world's financial markets and, had not emergency measures been enacted, might have seriously damaged them. (Then again, there is a laissez-faire argument for letting the fund fail.)

I admit I take a certain self-serving pleasure from this story since my own escapades pale by comparison. It's not clear, however, that the LTCM collapse was the fault of the Nobel laureates and their models. Many believe it was a consequence of a "perfect storm" in the markets, a vanishingly unlikely confluence of chance events. (The claim that Merton and Scholes were not implicated is nevertheless a bit disingenuous, since many invested in LTCM precisely because the fund was touting them and their models.)

The specific problems encountered by LTCM concerned a lack of liquidity in world markets, and this was exacerbated by the disguised dependence of a number of factors that were assumed to be independent. Consider, for illustration's sake, the likelihood that 3,000 specific people will die in New York on any given day. Provided that there is no connection among them, this is an impossibly minuscule number—a small probability raised to the 3,000th power. If most of the people work in a pair of buildings, however, the independence assumption that allows us to multiply probabilities fails. The 3,000 deaths are still extraordinarily unlikely, but not impossibly minuscule. Of course, the probabilities associated with possible LTCM scenarios were nowhere near as small and, according to some, could and should have been anticipated.

Are Insider Trading and Stock Manipulation So Bad?

It's natural to take a moralistic stance toward the corporate fraud and excess that have dominated business news the last couple of years. Certainly that attitude has not been completely absent from this book. An elementary probability puzzle and its extensions suggest, however, that some arguments against insider trading and stock manipulation are rather weak. Moral outrage, rather than actual harm to investors, seems to be the primary source of many people's revulsion toward these practices.

Let me start with the original puzzle. Which of the following two situations would you prefer to be in? In the first one you're given a fair coin to flip and are told that you will receive $1,000 if it lands heads and lose $1,000 if it lands tails. In the second you're given a very biased coin to flip and must decide whether to bet on heads or tails. If it lands the way you predict you win $1,000 and, if not, you lose $1,000. Although most people prefer to flip the fair coin, your chances of winning are 1/2 in both situations, since you're as likely to pick the biased coin's good side as its bad side.

Consider now a similar pair of situations. In the first one you are told you must pick a ball at random from an urn containing 10 green balls and 10 red balls. If you pick a green one, you win $1,000, whereas if you pick a red one, you lose $1,000. In the second, someone you thoroughly distrust places an indeterminate number of green and red balls in the urn. You must decide whether to bet on green or red and then choose a ball at random. If you choose the color you bet on, you win $1,000 and, if not, you lose $1,000. Again, your chances of winning are 1/2 in both situations.

Finally, consider a third pair of similar situations. In the first one you buy a stock that is being sold in a perfectly efficient market and your earnings are $1,000 if it rises the next day and –$1,000 if it falls. (Assume that in the short run it moves up with probability 1/2 and down with the same probability.) In the second there is insider trading and manipulation and the stock is very likely to rise or fall the next day as a result of these illegal actions. You must decide whether to buy or sell the stock. If you guess correctly, your earnings are

$1,000 and, if not, -$1,000. Once again your chances of winning are 1/2 in both situations. (They may even be slightly higher in the second situation since you might have knowledge of the insiders' motivations.)

In each of these pairs, the unfairness of the second situation is only apparent. You have the same chance of winning that you do in the first situation. I do not by any means defend insider trading and stock manipulation, which are wrong for many other reasons, but I do suggest that they are, in a sense, simply two among many unpredictable factors affecting the price of a stock.

I suspect that more than a few cases of insider trading and stock manipulation result in the miscreant guessing wrong about how the market will respond to his illegal actions. This must be depressing for the perpetrators (and funny for everyone else).

Expected Value, Not Value Expected

What can we anticipate? What should we expect? What's the likely high, low, and average value? Whether the quantity in question is height, weather, or personal income, extremes are more likely to make it into the headlines than are more informative averages. "Who makes the most money," for example, is generally more attention-grabbing than "what is the average income" (although both terms are always suspect because—surprise—like companies, people lie about how much money they make).

Even more informative than averages, however, are distributions. What, for example, is the distribution of all incomes and how spread out are they about the average? If the average income in a community is $100,000, this might reflect the fact that almost everyone makes somewhere between $80,000 and $120,000, or it might mean that a big majority earns less than $30,000 and shops at Kmart, whose spokesperson, the (too) maligned Martha Stewart, also lives in town and brings the average up to $100,000. "Expected value" and "standard deviation" are two mathematical notions that help clarify these issues.

An expected value is a special sort of average. Specifically, the expected value of a quantity is the average of its values, but weighted according to their probabilities. If, for example, based on analysts' recommendations, our own assessment, a mathematical model, or some other source of information, we assume that 1/2 of the time a stock will have a 6 percent rate of return, that 1/3 of the time it will have a –2 percent rate of return, and that the remaining 1/6 of the time it will have a 28 percent rate of return, then, on average, the stock's rate of return over any given six periods will be 6 percent three times, –2 percent twice, and 28 percent once. The expected value of its return is simply this probabilistically weighted average— (6% + 6% + 6% + (–2%) + (–2%) + 28%)/6, or 7%.

Rather than averaging directly, one generally obtains the expected value of a quantity by multiplying its possible values by their probabilities and then adding up these products. Thus $.06 \times 1/2 + (-.02) \times 1/3 + .28 \times 1/6 = .07$, or 7%, the expected value of the above stock's return. Note that the term "mean" and the Greek letter μ (mu) are used interchangeably with "expected value," so 7% is also the mean return, μ.

The notion of expected value clarifies a minor investing mystery. An analyst may simultaneously and without contradiction believe that a stock is very likely to do well but that, on average, it's a loser. Perhaps she estimates that the stock will rise 1 percent in the next month with probability 95 percent and that it will fall 60 percent in the same time period with probability 5 percent. (The probabilities might come, for example, from an appraisal of the likely outcome of an impending court decision.) The expected value of its price change is thus $(.01 \times .95) + (-.60) \times .05)$, which equals $-.021$ or an expected loss of 2.1%. The lesson is that the expected value, -2.1%, is not the value expected, which is 1%.

The same probabilities and price changes can also be used to illustrate two complementary trading strategies, one that usually results in small gains but sometimes in big losses, and one that usually results in small losses but sometimes in big gains. An investor who's willing to take a risk to regularly make some "easy money" might sell puts on the above stock, puts that expire in a month and whose strike price is a little under the present price. In effect, he's betting that the stock won't decline in the next month. Ninety-five percent of the time he'll be right, and he'll keep the put premiums and make a little money. Correspondingly, the buyer of the puts will lose a little money (the put premiums) 95 percent of the time. Assuming the probabilities are accurate, however, when the stock declines, it declines by 60 percent, and so the puts (the right to sell the stock at a little under the original price) become very valuable 5 percent of the time. The buyer of the puts then makes a lot of money and the seller loses a lot.

Investors can play the same game on a larger scale by buying and selling puts on the S&P 500, for example, rather than on any particular stock. The key to playing is coming up with reasonable probabilities for the possible returns, numbers about which people are as likely to differ as they are in their preferences for the above two strategies. Two exemplars of these two types of investor are Victor Niederhoffer, a well-known futures trader and author of *The Education of a Speculator*, who lost a fortune by selling puts a few years ago, and Nassim Taleb, another trader and the author of *Fooled by Randomness*, who makes his living by buying them.

For a more pedestrian illustration, consider an insurance company. From past experience, it has good reason to believe that each year, on average, one out of every 10,000 of its homeowners' policies will result in a claim of $400,000, one out of 1,000 policies will result in a claim of $60,000, one out of 50 will result in a claim of $4,000, and the remainder will result in a claim of $0. The insurance company would like to know what its average payout will be per policy written. The answer is the expected value, which in this case is ($400,000 × 1/10,000) + ($60,000 × 1/1,000) + ($4,000 × 1/50) + ($0 × 9,979/10,000) = $40 + $60 + $80 + $0 = $180. The premium the insurance company charges the homeowners will no doubt be at least $181.

Combining the techniques of probability theory with the definition of expected value allows for the calculation of more interesting quantities. The rules for the World Series of baseball, for example, stipulate that the series ends when one team wins four games. The rules further stipulate that team A plays in its home stadium for games 1 and 2 and however many of games 6 and 7 are necessary, whereas team B plays in its home stadium for games 3, 4, and, if necessary, game 5. If the teams are evenly matched, you might be interested in the expected number of games that will be played in each team's stadium. Skipping the calculation, I'll simply note that team A can expect to play 2.9375 games and team B 2.875 games in their respective home stadiums.

Almost any situation in which one can calculate (or reasonably estimate) the probabilities of the values of a quantity allows us to determine the expected value of that quantity. An example more tractable than the baseball problem concerns the decision whether to park in a lot or illegally on the street. If you park in a lot, the rate is $10 or $14, depending upon whether you stay for less than an hour, the probability of which you estimate to be 25 percent. You may, however, decide to park illegally on the street and have reason to believe that 20 percent of the time you will receive a simple parking ticket for $30, 5 percent of the time you will receive an obstruction of traffic citation for $100, and 75 percent of the time you will get off for free.

The expected value of parking in the lot is ($10 × .25) + ($14 × .75), which equals $13. The expected value of parking on the street is ($100 × .05) + ($30 × .20) + ($0 × .75), which equals $11. For those to whom this is not already Greek, we might say that μ_L, the mean costs of parking in the lot, and μ_S, the mean cost of parking on the street, are $13 and $11, respectively.

Even though parking in the street is cheaper on average (assuming money was your only consideration), the variability of what you'll have to pay there is much greater than it is with the lot. This brings us to the notion of standard deviation and stock risk.

What's Normal? Not Six Sigma

Risk in general is frightening, and the fear it engenders explains part of the appeal of quantifying it. Naming bogeymen tends to tame them, and chance is one of the most terrifying bogeyman around, at least for adults.

So how might one get at the notion of risk mathematically? Let's start with "variance," one of several mathematical terms for variability. Any chance-dependent quantity varies and deviates from its mean or average; it's sometimes more than the average, sometimes less. The actual temperature, for example, is sometimes warmer than the mean temperature, sometimes cooler. These deviations from the mean constitute risk and are what we want to quantify. They can be positive or negative, just as the actual temperature minus the mean temperature can be positive or negative, and hence they tend to cancel out. If we square them, however, the deviations are all positive, and we come to the definition: the variance of a chance-dependent quantity is the expected value of all its squared deviations from the mean. Before I numerically illustrate this, note the etymological/psychological association of risk with "deviation from the mean." This is a testament, I suspect, to our fear not only of risk but of anything unusual, peculiar, or deviant.

Be that as it may, let's switch from temperature back to our parking scenario. Recall that the mean cost of parking in the lot is $13, and so ($10 − $13)^2 and ($14 − $13)^2, which equal $9 and $1, respectively, are the squares of the deviations of the two possible costs from the mean. They don't occur equally frequently, however. The first occurs with probability 25%, and the second occurs with probability 75%, and so the variance, the expected value of these numbers, is ($9 × .25) + ($1 × .75), or $3. More commonly used in statistical applications in finance and elsewhere is the square root of the variance, which is usually symbolized by the Greek letter σ (sigma). Termed the "standard deviation," it is in this case the square root of $3, or approximately $1.73. The standard deviation is (not exactly, but can be thought of as) the average deviation from the mean, and it is the most common mathematical measure of risk.

Forget the numerical examples if you like, but remember that, for any quantity, the larger the standard deviation, the more spread out its possible values are about the mean; the smaller it is, the more tightly the possible values cluster around the mean. Thus, if you read that in Japan the standard deviation of personal incomes is much less than it is in the United States, you should infer that Japanese incomes vary considerably less than U.S. incomes.

Returning to the street, you may wonder what the variance and standard deviation are of your parking costs there. The mean cost of parking in the street is $11, and the squares of

the deviations of the three possible costs from the mean are
($100 − $11)², ($30 − $11)², and ($0 − $11)², or $7,921,
$361, and $121, respectively. The first occurs with probability
5%, the second with probability 20%, and the third with
probability 75%, and so the variance, the expected values of
these numbers, is ($7,921 × .05) + ($361 × .20) + ($121 ×
.75), or $559. The square root of this gives us the standard
deviation of $23.64, more than thirteen times the standard
deviation of parking in the lot.

Despite this blizzard of numbers, I reiterate that all we
have done is quantify the obvious fact that the possible outcomes of parking on the street are much more varied and unpredictable than those of parking in the lot. Even though the
average cost of parking in the street ($11) is less than that of
parking in the lot ($13), most would prefer to incur less risk
and would therefore park in the lot for prudential reasons, if
not moral ones.

This brings us to the market's use of standard deviation
(sigma) to measure a stock's volatility. Let's use the same approach to calculate the variance of the returns for our stock
that yields a rate of 6% about 1/2 the time, −2% about 1/3 of
the time, and 28% the remaining 1/6 of the time. The mean
or expected value of its returns is 7%, and so the squares of
the deviations from the mean are $(.06 − .07)^2$, $(−.02 − .07)^2$,
and $(.28 − .07)^2$ or .0001, .0081, and .0441, respectively.
These occur with probabilities 1/2, 1/3, and 1/6, and so the
variance, the expected value of the squares of these deviations
from the mean, is (.0001 × 1/2) + (.0081 × 1/3) + (.0441 ×
1/6), which is .01. The square root of .01 is .10 or 10%, and
this is the standard deviation of the returns for this stock.

The Greek lesson again: The expected value of a quantity is
its (probabilistically weighted) average and is symbolized by
the letter μ (mu), and the standard deviation of a quantity is a
measure of its variability and is symbolized by the letter

σ (sigma). If the quantity in question is the rate of return on a stock price, its volatility is generally taken to be the standard deviation.

If there are only two or three possible values a quantity might assume, the standard deviation is not that helpful a notion. It becomes very useful, however, when a quantity can assume many different values and these values, as they often do, have an approximately normal bell-shaped distribution—high in the middle and tapering off on the sides. In this case, the expected value is the high point of the distribution. Moreover, approximately 2/3 of the values (68 percent) lie within one standard deviation of the expected value, and 95 percent of the values lie within two standard deviations of the expected value.

Before we go on, let's list a few of the quantities that have a normal distribution: age-specific heights and weights, natural gas consumption in a city for any given winter day, water use between 2 A.M. and 3 A.M. in a given city, thicknesses of a particular machined part coming off an assembly line, I.Q.s (whatever it is that they measure), the number of admissions to a large hospital on any given day, distances of darts from a bull's-eye, leaf sizes, nose sizes, the number of raisins in boxes of breakfast cereal, and possible rates of return for a stock. If we were to graph any of these quantities, we would obtain bell-shaped curves whose values are clustered about the mean.

Take as an example the number of raisins in a large box of cereal. If the expected number of raisins is 142 and the standard deviation is 8, then the high point of the bell-shaped graph would be at 142. About two-thirds of the boxes would contain between 134 and 150 raisins, and 95 percent of the boxes would contain between 126 and 158 raisins.

Or consider the rate of return of a conservative stock. If the possible rates are normally distributed with an expected value of 5.4 percent and a volatility (standard deviation, that is) of only 3.2 percent, then about two-thirds of the time, the rate of return will be between 2.2 percent and 8.6 percent, and 95 percent of the time the rate will be between −1 percent and 11.8 percent. You might prefer this stock to a more risky one with the same expected value but a volatility of, say, 20.2 percent. About two-thirds of the time, the rate of return of this more volatile stock will be between −14.8 percent and 25.6 percent, and 95 percent of the time it will be between −35 percent and 45.8 percent.

In all cases, the more standard deviations from the expected value, the more unusual the result. This fact helps account for the many popular books on management and quality control having the words "six sigma" in their titles. The covers of many of these books suggest that by following their precepts, you can attain results that are six standard deviations above the norm, leading, for example, to a minuscule number of product defects. A six-sigma performance is, in fact, so unlikely that the tables in most statistics texts don't even include values for it. If you look into the books on management, however, you learn that Sigma is usually capitalized and means something other than sigma, the standard deviation of a chance-dependent quantity. A new oxymoron: minor capital offense.

Whether they are defects, nose sizes, raisins, or water use in a city, almost all normally distributed quantities can be thought of as the average or sum of many factors (genetic, physical, social, or financial). This is not an accident: The so-called Central Limit Theorem states that averages and sums of a sufficient number of chance-dependent quantities are always normally distributed.

As we'll see in chapter 8, however, not everyone believes that stocks' rates of return are normally distributed.

7 | Diversifying Stock Portfolios

*L*ong before my children's fascination with Super Mario Brothers, Tetris, and more recent addictive games, I spent interminable hours as a kid playing antediluvian, low-tech Monopoly with my two brothers. The game requires the players to roll dice and move around the board buying, selling, and trading real estate properties. Although I paid attention to the probabilities and expected values associated with various moves (but not to what have come to be called the game's Markov chain properties), my strategy was simple: Play aggressively, buy every property whether it made sense or not, and then bargain to get a monopoly. I always traded away railroads and utilities if I could, much preferring to build hotels on the real estate I owned instead.

A Reminiscence and a Parable

Although the game's get-out-of-jail-free card was one of the few ties to the present-day stock market, I've recently had a tiny epiphany. On some atavistic level I've likened hotel building to stock buying and the railroads and utilities to bonds. Railroads and utilities seemed safe in the short run, but the ostensibly risky course of putting most of one's money into building hotels was ultimately more likely to make one a winner (especially since we occasionally altered the rules to allow unlimited hotel building on a property).

Was my excessive investment in WorldCom a result of a bad generalization from playing Monopoly? I strongly doubt it, but such just-so stories come naturally to mind. Aside from the jail card, a board game called WorldCom would have few features in common with Monopoly (but might more closely resemble Grand Theft Auto). Different squares along players' paths would call for SEC investigations, Eliot Spitzer prosecutions, IPO giveaways, or favorable analyst ratings. If you attained CEO status, you would be allowed to borrow up to $400 million ($1 billion in later versions of the game), whereas if you were reduced to the rank of employee, you would have to pay a coffee fee after each move and invest a certain portion of your savings in company stock. If you were unfortunate enough to become a stockholder, you would be required to remove your shirt while playing, while if you became CFO, you would receive stock options and get to keep the stockholders' shirts. The object of the game would be to make as much money and collect as many of your fellow players' shirts as possible before the company went bankrupt.

The game might be fun with play money; it wasn't with the real thing.

Here's a better analogue for the market. People are milling around a huge labyrinthine bazaar. Occasionally some of the booths in the bazaar attract a swarm of people jostling to buy their wares. Likewise, some booths are occasionally devoid of any prospective customers. At any given time most booths have a few customers. At the intersections of the bazaar's alleys are sales people from some of the bigger booths as well as well-traveled seers. They know the various sections of the bazaar intimately and claim to be able to foretell the fortunes of various booths and collections of booths. Some of these sales people and some of the prognosticators have very large bullhorns and can be heard throughout the bazaar, while others make do by shouting.

In this rather primitive setting, many aspects of the stock market can already be discerned. The forebears of technical traders might be those who buy from booths where crowds are developing, while the forebears of fundamental traders might be those who coolly weigh the worth of the goods on display. The seers are the progenitors of analysts, the sales people progenitors of brokers. The bullhorns are a rudimentary form of business media, and, of course, the goods on sale are companies' stocks. Crooks and swindlers have their ancestors as well with some of the booths hiding their shoddy merchandise under the better goods.

If everyone, not just the booth owners, could sell as well as buy, this would be a better elemental model of an equities market. (I don't intend this as an historical account, but merely as an idealized narrative.) Nevertheless, I think it's clear that stock exchanges are natural economic phenomena. It's not hard to imagine early analogues of options trading, corporate bonds, or diversified holdings developing out of such a bazaar.

Maybe there'd even be some arithmeticians around too, analyzing booths' sales and devising purchasing strategies. In acting on their theories, some might even lose their togas and protractors.

Are Stocks Less Risky Than Bonds?

Perhaps because of Monopoly, certainly because of WorldCom, and for many other reasons, the focus of this book has been the stock market, not the bond market (or real estate, commodities, and other worthy investments). Stocks are, of course, shares of ownership in a company, whereas bonds are loans to a company or government, and "everybody knows" that bonds are generally safer and less volatile than stocks, although the latter have a higher rate of return. In fact, as Jeremy Siegel reports in *Stocks for the Long Run*, the average annual rate of return for stocks between 1802 and 1997 was 8.4 percent; the rate on treasury bills over the same period was between 4 percent and 5 percent. (The rates that follow are before inflation. What's needless to say, I hope, is that an 8 percent rate of return in a year of 15 percent inflation is much worse than a 4 percent return in a year of 3 percent inflation.)

Despite what "everybody knows," Siegel argues in his book that, as with Monopoly's hotels and railroads, stocks are actually less risky than bonds because, over the long run, they have performed so much better than bonds or treasury bills. In fact, the longer the run, the more likely this has been the case. (Comments like "everybody knows" or "they're all doing this" or "everyone's buying that" usually make me itch. My background in mathematical logic has made it difficult for me to interpret "all" as signifying something other than all.) "Everybody" does have a point, however. How can we believe Siegel's claims, given that the standard deviation for stocks' annual rate of return has been 17.5 percent?

If we assume a normal distribution and allow ourselves to get numerical for a couple of paragraphs, we can see how stomach-churning this volatility is. It means that about two-thirds of the time, the rate of return will be between –9.1 percent and 25.9 percent (that is, 8.4 percent plus or minus 17.5 percent), and about 95 percent of the time the rate will be between –26.6 percent and 43.4 percent (that is, 8.4 percent plus or minus two times 17.5 percent). Although the precision of these figures is absurd, one consequence of the last assertion is that the returns will be worse than –26.6 percent about 2.5 percent of the time (and better than 43.4 percent with the same frequency). So about once every forty years (1/40 is 2.5 percent), you will lose more than a quarter of the value of your stock investments and much more frequently than that do considerably worse than treasury bills.

These numbers certainly don't seem to indicate that stocks are less risky than bonds over the long term. The statistical warrant for Siegel's contention, however, is that over time, the returns even out and the deviations shrink. Specifically, the annualized standard deviation for rates of return over a number N of years is the standard deviation divided by the square root of N. The larger N is, the smaller is the standard deviation. (The cumulative standard deviation is, however, greater.) Thus over any given four-year period the annualized standard deviation for stock returns is 17.5%/2, or 8.75%. Likewise, since the square root of 30 is about 5.5, the annualized standard deviation of stock returns over any given thirty-year period is only 17.5%/5.5, or 3.2%. (Note that this annualized thirty-year standard deviation is the same as the annual standard deviation for the conservative stock mentioned in the example at the end of chapter 6.)

Despite the impressive historical evidence, there is no guarantee that stocks will continue to outperform bonds. If you look at the period from 1982 to 1997, the average annual rate of return for stocks was 16.7 percent with a standard deviation of 13.1 percent, while the returns for bonds were between 8 percent and 9 percent. But from 1966 to 1981, the average annual rate of return for stocks was 6.6 percent with a standard deviation of 19.5 percent, while the returns for bonds were about 7 percent.

So is it really the case that, despite the debacles, deadbeats, and doomsday equities like WCOM and Enron, the less risky long-term investment is in stocks? Not surprisingly, there is a counterargument. Despite their volatility, stocks as a whole have proven less risky than bonds over the long run because their average rates of return have been considerably higher. Their rates of return have been higher because their prices have been relatively low. And their prices have been relatively low because they've been viewed as risky and people need some inducement to make risky investments.

But what happens if investors believe Siegel and others, and no longer view stocks as risky? Then their prices will rise because risk-averse investors will need less inducement to buy them; the "equity-risk premium," the amount by which stock returns must exceed bond returns to attract investors, will decline. And the rates of return will fall because prices will be higher. And stocks will therefore be riskier because of their lower returns.

Viewed as less risky, stocks become risky; viewed as risky, they become less risky. This is yet another instance of the skittish, self-reflective, self-corrective dynamic of the market. Interestingly, Robert Shiller, a personal friend of Siegel, looks at the data and sees considerably lower stock returns for the next ten years.

Market practitioners as well as academics disagree. In early October 2002, I attended a debate between Larry Kudlow, a CNBC commentator and Wall Street fixture, and Bob Prechter, a technical analyst and Elliot wave proponent. The audience at the CUNY graduate center in New York seemed affluent and well-educated, and the speakers both seemed very sure of themselves and their predictions. Neither seemed at all affected by the other's diametrically opposed expectations. Prechter anticipated very steep declines in the market, while Kudlow was quite bullish. Unlike Siegel and Shiller, they didn't engage on any particulars and generally talked past each other.

What I find odd about such encounters is how typical they are of market discussions. People with impressive credentials regularly expatiate upon stocks and bonds and come to conclusions contrary to those of other people with equally impressive credentials. An article in the *New York Times* in November 2002 is another case in point. It described three plausible prognoses for the market—bad, so-so, and good—put forth by economic analysts Steven H. East, Charles Pradilla, and Abby Joseph Cohen, respectively. Such stark disagreement happens very rarely in physics or mathematics. (I'm not counting crackpots who sometimes receive a lot of publicity but aren't taken seriously by anybody knowledgeable.)

The market's future course may lie beyond what, in chapter 9, I term the "complexity horizon." Nevertheless, aside from some real estate, I remain fully vested in stocks, which may or may not result in my remaining fully shirted.

The St. Petersburg Paradox and Utility

Reality, like the perfectly ordinary woman in Virginia Woolf's famous essay "Mr. Bennett and Mrs. Brown," is endlessly complex and impossible to capture completely in any model. Expected value and standard deviation seem to reflect the ordinary meanings of average and variability most of the time, but it's not hard to find important situations where they don't.

One such case is illustrated by the so-called St. Petersburg paradox. It takes the form of a game that requires that you flip a coin repeatedly until a tail first appears. If a tail appears on the first flip, you win \$2. If the first tail appears on the second flip, you win \$4. If the first tail appears on the third flip, you win \$8, and, in general, if the first tail appears on the Nth flip, you win 2^N dollars. How much would you be willing to pay to play this game? One could argue that you should be willing to pay any amount to play this game.

To see why this is so, recall that the probability of a sequence of independent events such as coin flips is obtained by multiplying the probabilities of each of the events. Thus the probability of getting the first tail, T, on the first flip is 1/2; of getting a head and then the first tail on the second flip, HT, is $(1/2)^2$ or 1/4; of getting the first tail on the third flip, HHT, is $(1/2)^3$ or 1/8; and so on. Putting these probabilities and the possible winnings associated with them into the formula for expected value, we see that the expected value of the game is ($2 × 1/2) + ($4 × 1/4) + ($8 × 1/8) + ($16 × 1/16) + ... $(2^N × (1/2))^N$ + ... All of these products are 1, there are infinitely many of them, and so their sum is infinite. The failure of expected value to capture our intuitions becomes clear when you ask yourself why you'd be reluctant to pay even a measly $1,000 for the privilege of playing this game.

The most common resolution is roughly that provided by the eighteenth century mathematician Daniel Bernoulli, who wrote that people's enjoyment of any increase in wealth (or regret at any decrease) is "inversely proportionate to the quantity of goods previously possessed." The fewer dollars you have, the more you appreciate gaining one and the more you fear losing one, and so, for almost everyone, the likely prospect of losing $1,000 more than cancels the remote possibility that you'll win, say, a billion dollars.

What's important is the "utility" to you of the dollars that you receive, and this utility drops off as you receive more of them. (Note that this is not irrelevant to the rationale for progressive taxation.) For this reason people consider not the dollar amount involved in any investment (or game), but the utility of the dollar amount for the individual involved. The St. Petersburg paradox disappears, for example, if we consider a so-called logarithmic utility function, which attempts to reflect the slowly diminishing satisfaction of having more money and which results in the expected value of the game above being finite. Other versions of the game, in which the payoffs increase even faster, require even slower-growing utility functions so that the expected value remains finite.

People do differ in their utility assignments. Some are so acquisitive that the 741,783,219th dollar is almost as dear to them as the first; others are so laid back that their 25,000th dollar is almost worthless to them. There are probably relatively few of the latter, although my father in his later years came close. His attitude suggests that utility functions vary not only across people but also over time. Furthermore, utility may not be so easily described by simple functions since, for example, there may be variations in the utility of money as one approaches a certain age or reaches some financial milestone such as X million dollars. And we're back to Virginia Woolf's essay.

Portfolios: Benefiting from the Hatfields and McCoys

John Maynard Keynes wrote, "Practical men, who believe themselves to be quite exempt from any intellectual influences, are usually the slaves of some defunct economist. Madmen in authority, who hear voices in the air, are distilling their frenzy from some academic scribbler of a few years back." A corollary of this is that fund managers and stock gurus, who slickly dispense their investment ideas and advice, generally derive them from a previous generation's Nobel prize-winning finance professor.

To get a taste of what a couple more of these Nobelists have written, assume you're a fund manager intent on measuring the expected return and volatility (risk) of a portfolio. In stock market contexts a portfolio is simply a collection of different stocks—a mutual fund, for example, or Uncle Jake's ragbag of mysterious picks, or a nightmare inheritance containing a bunch of different stocks, all in telecommunications. Portfolios like the latter that are so lacking in diversification often become portfolios lacking in dollars. How can you more judiciously choose stocks to maximize a portfolio's returns and minimize its risks?

Let's first envision a simple portfolio consisting of only three stocks, Abbey Roads, Barkley Hoops, and Consolidated Fragments. Let's further assume that 40 percent (or $40,000) of a $100,000 portfolio is in Abbey, 25 percent in Barkley, and the remaining 35 percent in Consolidated. Assume further that the expected rate of return from Abbey is 8 percent, from Barkley is 13 percent, and from Consolidated is 7 percent. Using these weights, we compute that the expected return from the portfolio as a whole is $(.40 \times .08) + (.25 \times .13) + (.35 \times .07)$, which is .089 or 8.9 percent.

Why not put all our money in Barkley Hoops since its expected rate of return is the highest of the three stocks? The answer has to do with volatility and the risk of not diversifying, of putting all one's proverbial eggs in one basket. (The result, as was the case with my WorldCom misadventure, may well be egg on one's face and the transformation of one's nest egg into a scrambled egg if not a goose egg. Sorry, but thought of the stock even now sometimes momentarily unhinges me.) If you were indifferent to risk, however, and simply wanted to maximize your returns, you might well put all your money in Barkley Hoops.

So how does one determine the volatility—that is, sigma, the standard deviation—of a portfolio? Does one just weight the volatilities of the companies' stocks as we weighted their returns to get the volatility of the portfolio? In general, we can't do this because the stocks' performances are sometimes not independent of each other. When one goes up in response to some news, the others' chances of going up or down may be affected and this in turn affects their joint volatility.

Let me illustrate with an even simpler portfolio consisting of only two stocks, Hatfield Enterprises and McCoy Productions. They both produce thingamajigs, but history tells us that when one does well, the other suffers and vice versa, and that overall dominance seems to shift regularly back and forth between them. Perhaps Hatfield produces snow shovels and McCoy makes tanning lotion. To be specific, let's say that half the time Hatfield's rate of return is 40 percent and half the time it is –20 percent, so its expected rate of return is (.50 × .40) + (.50 × (–.20)), which is .10 or 10 percent. McCoy's returns are the same, but again it does well when Hatfield does poorly and vice versa.

The volatility of each company is the same too. Recalling the definition, we first find the squares of the deviations from the mean of 10 percent, or .10. These squares are $(.40 - .10)^2$ and $(-.20 - .10)^2$ or .09 and .09. Since they each occur half the time, the variance is (.50 × .09) + (.50 × .09), which is .09. The square root of this is .3 or 30 percent, which is the standard deviation or volatility of each company's returns.

But what if we don't choose one or the other to invest in, but split our investment funds and buy half as much of each stock? Then we're always earning 40 percent from half our investment and losing 20 percent on the other half, and our expected return is still 10 percent. But notice that this 10 percent return is constant. The volatility of the portfolio is zero! The reason is that the returns of these two stocks are not independent, but are perfectly negatively correlated. We get the same average return as if we bought either the Hatfield or the McCoy stock, but with no risk. This is a good thing; we get richer and don't have to worry about who's winning the battle between the Hatfields and the McCoys.

Of course, it's difficult to find stocks that are perfectly negatively correlated, but that is not required. As long as they aren't perfectly positively correlated, the stocks in a portfolio will decrease volatility somewhat. Even a portfolio of stocks from the same sector will be less volatile than the individual stocks in it, while a portfolio consisting of Wal-Mart, Pfizer, General Electric, Exxon, and Citigroup, the biggest stocks in their respective sectors, will provide considerably more protection against volatility. To find the volatility of a portfolio in general, we need what is called the "covariance" (closely related to the correlation coefficient) between any pair of stocks X and Y in the portfolio. The covariance between two stocks is roughly the degree to which they vary together—the degree, that is, to which a change in one is proportional to a change in the other.

Note that unlike many other contexts in which the distinction between covariance (or, more familiarly, correlation) and causation is underlined, the market generally doesn't care much about it. If an increase in the price of ice cream stocks is correlated to an increase in the price of lawn mower stocks, few ask whether the association is causal or not. The aim is to use the association, not understand it—to be right about the market, not necessarily to be right for the right reasons.

Given the above distinction, some of you may wish to skip the next three paragraphs on the calculation of covariance. Go directly to "For example, if we let H be the cost . . . ."

Technically, the covariance is the expected value of the product of the deviation from the mean of one of the stocks and the deviation from the mean of the other stock. That is, the covariance is the expected value of the product $[(X - \mu_X) \times (Y - \mu_Y)]$, where μ_X and μ_Y are the means of X and Y, respectively. Thus, if the stocks vary together, when the price of one is up, the price of the other is likely to be up too, so both deviations from the mean will be positive, and their product will be positive. And when the price of one is down, the price of the other is likely to be down too, so both deviations will be negative, and their product will again be positive. If the stocks vary inversely, however, when the price of one is up (or down), the price of the other is likely to be down (or up), so when the deviation of one stock is positive, that of the other is negative, and the product will be negative. In general and in short, we want negative covariance.

We may now use this notion of covariance to find the *variance* of a two-equity portfolio, p percent of which is in stock X and q percent in stock Y. The mathematics involves nothing more than squaring the sum of two terms. (Remember, however, that $(A + B)^2 = A^2 + B^2 + 2AB$.) By definition, the variance of the portfolio, $(pX + qY)$, is the expected value of the squares of its deviations from its mean, $p\mu_X + q\mu_Y$. That is, the variance of $(pX + qY)$ is the expected value of $[(pX + qY) - (p\mu_X + q\mu_Y)]^2$, which, upon rewriting, is the expected value of $[(pX - p\mu_X) + (qY - q\mu_Y)]^2$, which, using the algebra rule cited above, is the expected value of $[(pX - p\mu_X)^2 + (qY - q\mu_Y)^2 + 2 \times$ the expected value of $[(pX - p\mu_X) \times (qY - q\mu_Y)]$.

Minding (that is, factoring out) our p's and q's, we find that the variance of the portfolio, $(pX + qY)$, equals $[(p^2 \times$ the variance of X) + ($q^2 \times$ the variance of Y) + ($2pq \times$ the covariance of X and Y)]. If the stocks vary negatively (that is, have negative covariance), the variance of the portfolio is reduced by the last factor. (In the case of the Hatfield and McCoy stocks, the variance was reduced to zero.) And when they vary positively (that is, have positive covariance), the variance of the portfolio is increased by the last factor, a situation we want to avoid, volatility and risk being bad for our peace of mind and stomach.

For example, if we let H be the cost of a randomly selected homeowner's house in a given community and I be his or her household income, then the variance of $(H + I)$ is greater than the variance of H plus the variance of I. People who live in expensive houses generally have higher incomes than people who don't, so the extremes of the sum, house cost plus personal income, are going to be considerably greater than they would be if house cost and personal income did not have a positive covariance.

Likewise, if C is the number of classes skipped during the year by a randomly selected student in a large lecture and S is his score on the final exam, then the variance of (C + S) is smaller than the variance of C plus the variance of S. Students who miss a lot of classes generally (although certainly not always) achieve a lower score, so the extremes of the sum, number of classes missed plus exam scores, are going to be considerably less than they would be if number of classes missed and exam scores did not have a negative covariance.

When choosing stocks for a diversified portfolio, investors, as noted, generally look for negative covariances. They want to own equities like the Hatfield and the McCoy stocks and not like WCOM, say, and some other telecommunications stock. With three or more stocks in a portfolio, one uses the stocks' weights in the portfolio as well as the definitions just discussed to compute the portfolio's variance and standard deviation. (The algebra is tedious, but easy.) Unfortunately, the covariances between all possible pairs of stocks in the portfolio are needed for the computation, but good software, troves of stock data, and fast computers allow investors to determine a portfolio's risk (volatility, standard deviation) fairly quickly. With care, you can minimize the risk of a portfolio without hurting its expected rate of return.

Diversification and Politically Incorrect Funds

There are countless mutual funds, and many commentators have noted that there are more funds than there are stocks, as if this were a surprising fact. It isn't. In mathematical terms a fund is simply a set of stocks, so, theoretically at least, there are vastly more possible funds than there are stocks. Any set of n stocks (people, books, CDs) has 2^N subsets. Thus, if there were only 20 stocks in the world, there would be 2^{20} or approximately 1 million possible subsets of these stocks—1 million possible mutual funds. Of course, most of these subsets would not have a compelling reason for existence. Something more is needed, and that is the financial balancing act that ensures diversification and low volatility.

We can increase the number of possibilities even further by extending the notion of diversification. Instead of searching for individual stocks or whole sectors that are negatively correlated, we can search for concerns of ours that are negatively correlated. Say, for example, financial and social ones. A number of portfolios purport to be socially progressive and politically correct, but in general their performance is not stellar. Less appealing to many are funds that are socially regressive and politically incorrect but that do perform well. In this latter category many people would place tobacco, alcohol, defense contractors, fast food, or any of several others.

The existence of these politically incorrect funds suggests, for those passionately committed to various causes, a nonstandard strategy that exploits the negative correlation that sometimes exists between financial and social interests. Invest heavily in funds holding shares in companies that you find distasteful. If these funds do well, you make money, money that you could, if you wished, contribute to the political causes you favor. If these funds cool off, you can rejoice that the companies are no longer thriving, and your psychic returns will soar.

Such "diversification" has many applications. People often work for organizations, for example, whose goals or products they find unappealing and use part of their salary to counter the organization's goals or products. Taken to its extreme, diversification is something we do naturally in dealing with the inevitable trade-offs in our daily lives.

Of course, extending the notion of diversification to these other realms is difficult for several reasons. One is that quantifying contributions and payoffs is problematic. How do you place a numerical value on your efforts and their various consequences? The number of possible "funds," subsets of all your possible concerns, also grows exponentially.

Another problem derives from the logic of the notion of diversification. It often makes sense in life, where some combination of work, play, family, personal experiences, study, friends, money, and so forth, seems more likely to lead to satisfaction than, say, all toil or pure hedonism. Nevertheless, diversification may not be appropriate when you are trying to have a personal impact. Take charity, for example.

As the economist Steven Landsburg has argued, you diversify when investing to protect yourself, but when contributing to large charities in which your contributions are a small fraction of the total, your goal is presumably to help as much as possible. Since you incur no personal risk, if you truly think that Mothers Against Drunk Driving is more worthy than the American Cancer Society or the American Heart Association, why would you split your charitable dollars among them? The point isn't to insure that your money will do some good, but to maximize the good it will do. There are other situations too where bulleting one's efforts is preferable to a bland diversification.

Metaphorical extensions of the notion of diversification can be useful, but uncritical use of them can lead you to, in the words of W. H. Auden, "commit a social science."

Beta—Is It Better?

Returning to more quantitative matters, we choose stocks so that when some are down, others are up (or at least not as down), giving us a healthy rate of return with as little risk as possible. More precisely, given any portfolio of stocks, we grind the numbers describing their past performances and come up with estimates for their expected returns, volatilities, and covariances, and then use these to determine the expected returns and volatilities of the portfolio as a whole. We could, if we had the time, the price data, and fast computers, do this for a variety of different portfolios. The Nobel prize-winning economist Harry Markowitz, one of the originators of this approach, developed mathematical techniques for carrying out these calculations in the early 1950s, graphed his results for a few portfolios (computers weren't fast enough to do much more then), and defined what he called the "efficient frontier" of portfolios.

If we were to use these techniques and construct comparable graphs for a wide variety of contemporary portfolios, what would we find? Arraying the (degree of) volatility of these portfolios along the graph's horizontal axis and their expected rates of return along its vertical axis, we would see a swarm of points. Each point would represent a portfolio whose coordinates would be its volatility and expected return, respectively. We'd also notice that among all the portfolios having a given level of risk (that is, volatility, standard deviation), there would be one with the highest expected rate of return. If we single out the portfolio with the highest expected rate of return for each level of risk, we would obtain a curve, Markowitz's efficient frontier of optimal portfolios.

The more risky a portfolio on the efficient frontier curve is, the higher is its expected return. In part, this is because most investors are risk-averse, making risky stocks cheaper. The idea is that investors decide upon a risk level with which they're comfortable and then choose the portfolio with this risk level that has the highest possible return. Call this Variation One of the theory of portfolio selection.

Don't let this mathematical formulation blind you to the generality of the psychological phenomenon. Automobile engineers have noted, for example, that safety advances in automobile design (say anti-lock brakes) often result in people driving faster and turning more sharply. Their driving performance is enhanced rather than their safety. Apparently, people choose a risk level with which they're comfortable and then seek the highest possible return (performance) for it.

Inspired by this trade-off between risk and return, William Sharpe proposed in the 1960s what is now a common measure of the performance of a portfolio. It is defined as the ratio of the excess return of a portfolio (the difference between its expected return and the return on a risk-free treasury bill) to the portfolio's volatility (standard deviation). A portfolio might have a hefty rate of return, but if the volatility the investor must endure to achieve this return is roller coasterish, the portfolio's Sharpe measure won't be very high. By contrast, a portfolio with a moderate rate of return but a less anxiety-inducing volatility will have a higher Sharpe measure.

There are many complications to portfolio selection theory. As the Sharpe measure suggests, an important one is the existence of risk-free investments, such as U.S. treasury bills. These pay a fixed rate of return and have essentially zero volatility. Investors can always invest in such risk-free assets and can borrow at the risk-free rates as well. Moreover, they can combine risk-free investment in treasury bills with a risky stock portfolio.

Variation Two of portfolio theory claims that there is one and only one optimal stock portfolio on the efficient frontier with the property that some combination of it and a risk-free investment (ignoring inflation) constitute a set of investments having the highest rates of return for any given level of risk. If you wish to incur no risk, you put all your money into treasury bills. If you're comfortable with risk, you put all your money into this optimal stock portfolio. Alternatively, if you want to divide your money between the two, you put p% into the risk-free treasury bills and (100 – p)% into the optimal risky stock portfolio for an expected rate of return of [p × (risk-free return) + (1 – p) × (stock portfolio)]. An investor can also invest more money than he has by borrowing at the risk-free rate and putting this borrowed money into the risky portfolio.

In this refinement of portfolio selection, all investors choose the same optimal stock portfolio and then adjust how much risk they're willing to take by increasing or decreasing the percentage, p, of their holdings that they put into risk-free treasury bills.

This is easier said than done. In both variations the required mathematical procedures put enormous pressure on one's computing facilities, since countless calculations must be performed regularly on new data. The expected returns, variances, and covariances are, after all, derived from their values in the recent past. If there are twenty stocks in a portfolio, we would need to compute the covariance of every possible pair of stocks, and there are (20 × 19)/2, or 190, such covariances. If there were fifty stocks, we'd need to compute (50 × 49)/2, or 1,225 covariances. Doing this for each of a wide class of portfolios is not possible without massive computational power.

As a way to avoid much of the computational burden of updating and computing all these covariances, efficient frontiers, and optimal risky portfolios, Sharpe, yet another Nobel Prize winner in economics, developed (with others) what's called the "single index model." This Variation Three relates a portfolio's rate of return not to that of all possible pairs of stocks in the portfolio, but simply to the change in some index representing the market as a whole. If your portfolio or stock is statistically determined to be relatively more volatile than the market as a whole, then changes in the market will bring about exaggerated changes in the stock or portfolio. If it is relatively less volatile than the market as a whole, then changes in the market will bring about attenuated changes in the stock or portfolio.

This brings us to the so-called Capital Asset Pricing Model, which maintains that the expected excess return on one's stock or portfolio (the difference between the expected return on the portfolio, R_p, and the return on risk-free treasury bills, R_f) is equal to the notorious beta, symbolized by β, multiplied by the expected excess return of the general market (the difference between the market's expected return, R_m, and the return on risk-free treasury bills, R_f). In algebraic terms: $(R_p - R_f) = \beta(R_m - R_f)$. Thus, if you can get a sure 4 percent on treasury bills and if the expected return on a broad market index fund is 10 percent and if the relative volatility, beta, of your portfolio is 1.5, then the portfolio's expected return is obtained by solving $(R_p - 4\%) = 1.5(10\% - 4\%)$, which yields 13 percent for R_p. A beta of 1.5 means that your stock or portfolio gains (or loses) an average of 1.5 percent for every 1 percent gain (or loss) in the market as a whole.

Betas for the stocks of high-tech companies like World-Com are often considerably more than 1, meaning that changes in the market, both up and down, are magnified. These stocks are more volatile and thus riskier. Betas for utility company stocks, by contrast, are often less than 1, which means that changes in the market are muted. If a company has a beta of .5, then its expected return is obtained by solving $(R_p - 4\%) = .5(10\% - 4\%)$, which yields 7% for R_p, the expected return on the portfolio. Note that for short-term treasury bills, whose returns don't vary at all, beta is 0. To reiterate: Beta quantifies the degree to which a stock or a portfolio fluctuates in relation to market fluctuations. It is not the same as volatility.

This all sounds neat and clean, but you beta watch your step with all of these portfolio selection models. Specifically with regard to Variation Three, we might wonder where the number beta comes from. Who says your stock or portfolio will be 40 percent more volatile or 25 percent less volatile than the market as a whole? Here's the rough technique for finding beta. You check the change in the broad market for the last three months—say it's 3 percent—and check the change in the price of your stock or portfolio for the same period—say it's 4.1 percent. You do the same thing for the three months before that—say the numbers this time are 2 percent and 2.5 percent, respectively—and for the three months before that—say −1.2 percent and −3 percent, respectively. You continue doing this for a number of such periods and then on a graph you plot the points (3%, 4.1%), (2%, 2.5%), (−1.2%, −3%), and so on. Most of the time if you squint hard enough, you'll see a sort of linear relationship between changes in the market and changes in your stock or portfolio, and you then use standard mathematical methods for determining the line of closest fit through these points. The slope or steepness of this line is beta.

One problem with beta is that companies change over time, sometimes rather quickly. AT&T, for example, or IBM is not the same company it was twenty years ago or even two years ago. Why should we expect a company's relative volatility, beta, to remain the same? In the opposite direction is a related difficulty. Beta is often of very limited value in the short term and varies with the index chosen for comparison and the time period used in its definition. Still another problem is that beta depends on market returns, and market returns depend on a narrow definition of the market, namely just the stock market rather than stocks, bonds, real estate, and so forth. For all its limitations, however, beta can be a useful notion if it's not turned into a fetish.

You might compare beta to different people's emotional re-activity and expressiveness. Some respond to the slightest good news with outbursts of joy and to the tiniest hardship with wails of despair. At the other end of the emotional spec-trum are those who say "ouch" when they accidentally touch a scalding iron and allow themselves an "oh, good" when they win the lottery. The former have high emotional beta, the latter low emotional beta. A zero beta person would have to be unconscious, perhaps from ingesting too many beta-blockers. Unfortunate for the prospect of predicting the be-havior of people, however, is the commonplace that people's emotional betas vary depending on the type of stimulus a per-son faces. I'll leave out the examples, but this may be beta's biggest limitation as a measure of the relative volatility of a portfolio or stock. Betas may vary with the type of stimulus a company faces.

Whatever refinements of portfolio theory are developed, one salient point remains: Portfolios, although often less risky than individual stocks, are still risky (as millions of 401(k) returns attest). Some mathematical manipulation of the notions of variance and covariance and a few reasonable assumptions are sufficient to show that this risk can be partitioned into two parts. There is a systematic part that is related to general move-ments in the market, and there is a non-systematic part that is idiosyncratic to the stocks in the portfolio. The latter, non-systematic risk, specific to the individual stocks in the portfolio, can be eliminated or "diversified away" by an appropriate choice of thirty or so stocks. An irreducible core, however, re-mains inherent in the market and cannot be avoided. This sys-tematic risk depends on the beta of one's portfolio.

Or so the story goes. To the criticisms of beta above should be added the problems associated with forcing a non-linear world into a linear mold.

8 | Connectedness and Chaotic Price Movements

Near the end of my involvement with WorldCom, when I was particularly concerned about what the new day would bring, I would sometimes wake up very early, grab a Diet Coke, and check how the stock was faring on the German or English exchanges. As the computer was booting up, I grew more and more apprehensive. The European response to bad overnight news sometimes prefigured Wall Street's response, and I dreaded seeing a steeply downward-sloping graph pop up on my screen. More often the European exchanges treaded water on WCOM until trading began in New York. Occasionally I'd be encouraged when the stock was up there, but I soon learned that the small volume sold on overseas exchanges didn't always mean much.

Whether haunted by a bad investment or not, we're all connected. No investor is an island (or even a peninsula). Stated mathematically, this means that statistical independence often fails; your actions affect mine. Most accounts of the stock market acknowledge in a general way that we learn from and respond to one another, but a better understanding of the market requires that one's models reflect the complexity of investors' interaction. In a sense, the market *is* the interaction. Stocks R Us. Before discussing some of the consequences of this complexity, let me consider three such sources for it: one micro, one macro, and the third mucro (yup, it's a word).

The micro example involves insider trading, which has always struck me as an odd sort of crime. Few people who aren't psychopaths daydream about murder or burglary, but many investors, I suspect, fantasize about coming upon inside information and making a bundle from it. The thought of finding myself on a plane next to Bernie Ebbers and Jack Grubman (assuming they flew economy class on commercial airliners) and overhearing their conversation about an impending merger or IPO offering, for example, did cross my mind a few times. Insider trading seems the limit or culmination of what investors and traders do naturally: getting all the information possible and acting on it before others see and understand what they see and understand.

Insider Trading and
Subterranean Information Processing

The kind of insider trading I want to consider is relevant to seemingly unexplained price movements. It's also related to good poker playing, which may explain why the training program of at least one very successful hedge fund has a substantial unit on the game. The strategies associated with poker include learning not only the relevant probabilities but also the bluffing that is a necessary part of the game. Options traders often deal with relatively few other traders, many of whom they recognize, and this gives rise to the opportunity for feints, misdirection, and the exploitation of idiosyncrasies.

The example derives from the notion of common knowledge introduced in chapter 1. Recall that a bit of information is common knowledge among a group of people if they all know it, know that the others know it, know that the others know that they know it, and so on. Robert Aumann, who first defined the notion, proved a theorem that can be roughly paraphrased as follows: Two individuals cannot forever agree to disagree. As their beliefs, formed in rational response to different bits of private information, gradually become common knowledge, the beliefs change and eventually coincide.

When private information becomes common knowledge, it induces decisions and actions. As anyone who has overheard teenagers' gossip with its web of suppositions can attest, this transition to common knowledge sometimes relies on convoluted inferences about others' beliefs. Sergiu Hart, an economist at Hebrew University and one of a number of people who have built on Aumann's result, demonstrates this with an example relevant to the stock market. Superficially complicated, it nevertheless requires no particular background besides an ability to decode gossip, hearsay, and rumor and decide what others really think.

Hart asks us to consider a company that must make a decision. In keeping with the WorldCom leitmotif, let's suppose it to be a small telecommunications company that must decide whether to develop a new handheld device or a cell phone with a novel feature. Assume that the company is equally likely to decide on one or the other of these products, and assume further that whatever decision it makes, the product chosen has a 50 percent chance of being successful, say being bought in huge numbers by another company. Thus there are four equally likely outcomes: Handheld+, Handheld−, Phone+, Phone− (where Handheld+ means the handheld device was chosen for development and it was a success, Handheld− means the handheld was chosen but it turned out to be a failure, and similarly for Phone+ and Phone−).

Let's say there are two influential investors, Alice and Bob. They both decide that at the current stock price, if the chances of success of this product development are better than 50 percent, they should (continue to) buy, and if they're 50 percent or less they should (continue to) sell.

Furthermore, they are each privy to a different piece of information about the company. Because of her inside contacts, Alice knows which product decision was made, Handheld or Phone, but not whether it was successful or not.

Bob, because of his position with another company, stands to get the "rejects" from a failed phone project, so he knows whether or not the cell phone was chosen for development and failed. That is, Bob knows whether Phone– or not.

Let's assume that the handheld device was chosen for development. So the true situation is either Handheld+ or Handheld–. Alice therefore knows Handheld, while Bob knows that the decision is not Phone– (else he would have received the rejects).

After the first period (week, day, or hour), Alice sells since Handheld+ and Handheld– are equally likely, and one sells if the probability of success is 50 percent or less. Bob buys since he estimates that the probability for success is 2/3. With Phone– ruled out, the remaining possibilities are Handheld+, Handheld–, and Phone+, and two out of three of them are successes.

After the second period, it is common knowledge that the true situation is not Phone– since otherwise Bob would have sold in the first period. This is not news to Alice, who continues to sell. Bob continues to buy.

After the third period, it is common knowledge that it is not Phone (neither Phone+ nor Phone–) since otherwise Alice would have bought in the second period. Thus it's either Handheld+ or Handheld–. Both Bob and Alice take the probability of success to be 50 percent, thus both sell, and there is a mini-crash of the stock price. (Selling by both influential investors triggers a general sell-off.)

Note that at the beginning both Alice and Bob know that the true situation is not Phone–, but this knowledge is mutual, not common. Alice knows that Bob knows it is not Phone–, but Bob does not know that Alice knows this. From his position the true situation might be Phone+, in which case Alice would know Phone but not whether the situation is Phone+ or Phone–.

The example can be varied in a number of ways: there needn't be merely three periods before a crash, but an arbitrary number; there may be a bubble (sellers suddenly switching to become buyers) instead of a crash; there may be an arbitrarily large number of investors or investor groups; there may be an issue other than buying or selling under deliberation, perhaps a decision whether to employ one stock-picking approach rather than another.

In all these cases the stock's price can move in response to no external news. Nevertheless, the subterranean information processing leading to common knowledge among the investors eventually leads to precipitous and unexpected movement in the stock's price. Analysts will express surprise at the crash (or bubble) because "nothing happened."

The example is also relevant to what I suspect is a relatively common kind of insider trading, in which "partial insiders" are privy to bits of insider information but not to the whole story.

Trading Strategies, Whim, and Ant Behavior

A more macro-level interaction among investors occurs between technical traders and value traders. Also contributing over time to booms and busts, this interaction comes through clearly in computer models of the following commonsense dynamic.

Let's suppose that value traders perceive individual stocks or the market as a whole to be strongly undervalued. They start buying and, by doing so, raise prices. As prices increase, a trend develops and technical traders, as is their wont, follow it, increasing prices even further. Soon enough, the market is seen as overvalued by value traders, who begin to sell and thereby slow and then reverse the trend. The trend-following technical traders eventually follow suit, and the cycle begins over again. There are, of course, other sources of variation (one being the number of people who are technical traders and value traders at any given time), and the oscillations are irregular.

The bottom line of much of this modeling is that contrarian value traders have a stabilizing effect on the market, whereas technical traders increase volatility. So does computer-generated program trading, which tends to produce buying or selling in lockstep. There are other sorts of interaction among different classes of investors leading to cycles of varying duration, all of which have differential impacts on the others on which they are superimposed.

In addition to these more or less rational interactions among investors I must also note influences inspired by nothing more than whim, where behavior turns on a mucro. I recall many times, for example, reluctantly beginning work on a project when a niggling detail about some utterly irrelevant matter came to mind. It may have concerned the etymology of a word, or the colleague whose paper bag ripped open at a departmental meeting revealing an embarrassing magazine inside, or why caller ID misidentified a friend's telephone number. These in turn brought to mind the next in a train of associations and musings, which ultimately led me to an entirely different project. My impulsively deciding, while browsing in Borders, to make my first margin call on WCOM is another instance.

When this capriousness extends to influential analysts, the effect is more pronounced. In November 2002 the *New York Times* reported on such a case involving Jack Grubman, telecommunications analyst and anxious father. In an email to a friend Grubman allegedly stated that his boss, Sanford Weill, the chairman of Citigroup, helped get Grubman's children into an exclusive nursery school after he raised his rating of AT&T in 1999. Gretchen Morgenson, the article's author, further reported that Weill had his own personal reasons for wanting this upgrade. Whether these particular charges are true or not is immaterial. It's very hard to believe, however, that this sort of influence is rare.

Such episodes strongly suggest to me that there will never be a precise science of finance or economics. Buying and selling must surely partake of a similar iffiness, at least sometimes. *Butterfly Economics,* by the British economic theorist Paul Ormerod, faults these disciplines for not sufficiently taking into account the commonsense fact that people, whether knowledgeable or not, influence each other.

People do not, as chapter 2 demonstrated, have a set of fixed preferences on which they coolly and rationally base their economic decisions. The assumption that investors are sensitive only to price and a few ratios simplifies the mathematical models, but it is not always true to our experience of fads, fashions, and people's everyday monkey-see, monkey-do behavior.

Ormerod tells of an experiment involving not monkeys but ants that provides a useful metaphor. Two identical piles of food are set up at equal distances from a large nest of ants. Each pile is automatically replenished and the ants have no reason to prefer one to the other. Entomologists tell us that once an ant has found food, it usually returns to the same source. Upon returning to the nest, however, it physically stimulates other ants, who might be frequenting the other pile, to follow it to the first pile.

So where do the ants go? It might be speculated that either they would split into two roughly even groups or perhaps a large majority would arbitrarily settle on one or the other pile. Their actual behavior is counterintuitive. The number of ants going to each pile fluctuates wildly and doesn't ever settle down. A graph of these fluctuations looks suspiciously like a graph of the stock market.

And in a way, the ants are like stock traders (or people deciding whether or not to make a margin call). Upon leaving the nest, each ant must make a decision: Go to the pile visited last time, be influenced by another to switch piles, or switch piles of its own volition. This slight openness to the influence of other ants is enough to insure the complicated and volatile fluctuations in the number of ants visiting the two sites.

An astonishingly simple formal model of such influence is provided by Stephen Wolfram in his book *A New Kind of Science*. Imagine a colossally high brick wall wherein each brick rests on parts of two bricks below it and, except for the top row, has parts of two bricks above it. Imagine further that the top row has some red bricks and some green ones. The coloring of the bricks in the top row determines the coloring of the bricks in the second row as follows. Pick a brick in the second row and check the colors of the two bricks above it in the first row. If exactly one of these bricks is green, then the brick in the second row is colored green. If both or neither are green, then the brick is colored red. Do this for every brick in the second row.

The coloring of the bricks in the second row determines the coloring of the bricks in the third row in the same way, and in general, the coloring of the bricks in any row determines the coloring of the bricks in the row below it in the same way. That's it.

Now if we interpret a row of bricks as a collection of investors at any given instant, green ones for buyers and red ones for sellers, then the change from moment to moment of investor sentiment is reflected in the changing color composition of the succeeding rows of bricks. If we let P be the difference between the number of green bricks and the number of red bricks, then P is a rough analogue of a stock's price. Graphing it, we see that it oscillates up and down in a way that looks random.

The model can be made more realistic, but it is significant that even this bare-bones version, like the ant behavior, evinces a kind of internally generated random noise. This suggests that *part of* the oscillation of stock prices is also internally generated and is not a response to anything besides investors' reactions to each other. The theme of Wolfram's book, borne out here, is that complex behavior can result from very simple rules of interaction.

Chaos and Unpredictability

What is the relative importance of private information, investor trading strategies, and pure whim in predicting the market? What is the relative importance of conventional economic news (interest rates, budget deficits, accounting scandals, and trade balances), popular culture fads (in sports, movies, fashions), and germane political and military events (terrorism, elections, war) too disparate even to categorize? If we were to carefully define the problem, predicting the market with any precision is probably what mathematicians call a universal problem, meaning that a complete solution to it would lead immediately to solutions for a large class of other problems. It is, in other words, as hard a problem in social prediction as there is.

Certainly, too little notice is taken of the complicated connections among these variables, even the more clearly defined economic ones. Interest rates, for example, have an impact on unemployment rates, which in turn influence revenues; budget deficits affect trade deficits, which sway interest rates and exchange rates; corporate fraud influences consumer confidence, which may depress the stock market and alter other indices; natural business cycles of various periods are superimposed on one another; an increase in some quantity or index positively (or negatively) feeds back on another, reinforcing or weakening it and being reinforced or weakened in turn.

Few of these associations are accurately described by a straight-line graph and so they bring to a mathematician's mind the subject of nonlinear dynamics, more popularly known as chaos theory. The subject doesn't deal with anarchist treatises or surrealist manifestoes but with the behavior of so-called nonlinear systems. For our purposes these may be thought of as any collection of parts whose interactions and connections are described by nonlinear rules or equations. That is to say, the equations' variables may be multiplied together, raised to powers, and so on. As a consequence the system's parts are not necessarily linked in a proportional manner as they are, for example, in a bathroom scale or a thermometer; doubling the magnitude of one part will not double that of another—nor will outputs be proportional to inputs. Not surprisingly, trying to predict the precise long-term behavior of such systems is often futile.

Let me, in place of a technical definition of such nonlinear systems, describe instead a particular physical instance of one. Picture before you a billiards table. Imagine that approximately twenty-five round obstacles are securely fastened to its surface in some haphazard arrangement. You hire the best pool player you can find and ask him to place the ball at a particular spot on the table and take a shot toward one of the round obstacles. After he's done so, his challenge is to make exactly the same shot from the same spot with another ball.

Even if his angle on this second shot is off by the merest fraction of a degree, the trajectories of these two balls will very soon diverge considerably. An infinitesimal difference in the angle of impact will be magnified by successive hits of the obstacles. Soon one of the balls will hit an obstacle that the other misses entirely, at which point all similarity between the two trajectories ends.

The sensitivity of the billiard balls' paths to minuscule variations in their initial angles is characteristic of nonlinear systems. The divergence of the billiard balls is not unlike the disproportionate effect of seemingly inconsequential events, the missed planes, serendipitous meetings, and odd mistakes and links that shape and reshape our lives.

This sensitive dependence of nonlinear systems on even tiny differences in initial conditions is, I repeat, relevant to various aspects of the stock market in general, in particular its sometimes wildly disproportionate responses to seemingly small stimuli such as companies' falling a penny short of earnings estimates. Sometimes, of course, the differences are more substantial. Witness the notoriously large discrepancies between government economic figures on the size of budget surpluses and corporate accounting statements of earnings and the "real" numbers.

Aspects of investor behavior too can no doubt be better modeled by a nonlinear system than a linear one. This is so despite the fact that linear systems and models are much more robust, with small differences in initial conditions leading only to small differences in final outcomes. They're also easier to predict mathematically, and this is why they're so often employed whether their application is appropriate or not. The chestnut about the economist looking for his lost car keys under the street lamp comes to mind. "You probably lost them near the car," his companion remonstrates, to which the economist responds, "I know, but the light is better over here."

The "butterfly effect" is the term often used for the sensitive dependence of nonlinear systems, a characteristic that has been noted in phenomena ranging from fluid flow and heart fibrillations to epilepsy and price fluctuations. The name comes from the idea that a butterfly flapping its wings someplace in South America might be sufficient to change future weather systems, helping to bring about, say, a tornado in Oklahoma that would otherwise not have occurred. It also explains why long-range precise prediction of nonlinear systems isn't generally possible. This non-predictability is the result not of randomness but of complexity too great to fathom.

Yet another reason to suspect that parts of the market may be better modeled by nonlinear systems is that such systems' "trajectories" often follow a fractal course. The trajectories of these systems, of which the stock price movements may be considered a proxy, turn out to be aperiodic and unpredictable and, when examined closely, evince even more intricacy. Still closer inspection of the system's trajectories reveals yet smaller vortices and complications of the same general kind.

In general, fractals are curves, surfaces, or higher dimensional objects that contain more, but similar, complexity the closer one looks. A shoreline, to cite a classic example, has a characteristic jagged shape at whatever scale we draw it; that is, whether we use satellite photos to sketch the whole coast, map it on a fine scale by walking along some small section of it, or examine a few inches of it through a magnifying glass. The surface of the mountain looks roughly the same whether seen from a height of 200 feet by a giant or close up by an insect. The branching of a tree appears the same to us as it does to birds, or even to worms or fungi in the idealized limiting case of infinite branching.

As the mathematician Benoit Mandelbrot, the discoverer of fractals, has famously written, "Clouds are not spheres, mountains are not cones, coastlines are not circles, and bark is not smooth, nor does lightning travel in a straight line." These and many other shapes in nature are near fractals, having characteristic zigzags, push-pulls, bump-dents at almost every size scale, greater magnification yielding similar but ever more complicated convolutions.

And the bottom line, or, in this case, the bottom fractal, for stocks? By starting with the basic up-down-up and down-up-down patterns of a stock's possible movements, continually replacing each of these patterns' three segments with smaller versions of one of the basic patterns chosen at random, and then altering the spikiness of the patterns to reflect changes in the stock's volatility, Mandelbrot has constructed what he calls multifractal "forgeries." The forgeries are patterns of price movement whose general look is indistinguishable from that of real stock price movements. In contrast, more conventional assumptions about price movements, say those of a strict random-walk theorist, lead to patterns that are noticeably different from real price movements.

These multifractal patterns are so far merely descriptive, not predictive of specific price changes. In their modesty, as well as in their mathematical sophistication, they differ from the Elliott waves mentioned in chapter 3.

Even this does not prove that chaos (in the mathematical sense) reigns in (part of) the market, but it is clearly a bit more than suggestive. The occasional surges of extreme volatility that have always been a part of the market are not as nicely accounted for by traditional approaches to finance, approaches Mandelbrot compares to "theories of sea waves that forbid their swells to exceed six feet."

Extreme Price Movements, Power Laws, and the Web

Humans are a social species, which means we're all connected to each other, some in more ways than others. This is especially so in financial matters. Every investor responds not only to relatively objective economic considerations, but also in varying degrees to the pronouncements of national and world leaders (not least of those Mr. Greenspan), consumer confidence, analysts' ratings (bah), general and business media reports and their associated spin, investment newsletters, the behavior of funds and large institutions, the sentiments of friends, colleagues, and of course the much-derided brother-in-law.

The linkage of changes in stock prices to the varieties of investor responses and interactions suggests to me that communication networks, degrees of connectivity, and so-called small world phenomena ("Oh, you must know my uncle Waldo's third wife's botox specialist") can shine a light on the workings of Wall Street.

First the conventional story. Movements in a stock or index over small units of time are usually slightly positive or slightly negative, less frequently very positive or very negative. A large fraction of the time, the price will rise or fall between 0 percent and 1 percent; a smaller fraction of the time, it will rise or fall between 1 percent and 2 percent; a very small fraction of the time will the movement be more than, say, 10 percent up or down. In general, the movements are well described by a normal bell-shaped curve. The most likely change for a small unit of time is probably a minuscule jot above zero, reflecting the market's long-term (and recently invisible) upward bias, but the fact remains that extremely large price movements, whether positive or negative, are rare.

It's been clear for some time, however (that is, since Mandelbrot made it clear), that extreme movements are not as rare as the normal curve would predict. If you measure commodity price changes, for example, in each of a large number of small time units and make from these measurements a histogram, you will notice that the graph is roughly normal near its middle. The distribution of these price movements, however, seems to have "fatter tails" than the normal distribution, suggesting that crashes and bubbles in a stock, an index, or the entire market are less unlikely than many would like to admit. There is, in fact, some evidence that very large movements in stock prices are best described by a so-called power law (whose definition I'll get to shortly) rather than the tails of the normal curve.

An oblique approach to such evidence is via the notions of connectivity and networks. Everyone's heard people exclaim about how amazed they were to run into someone they knew so far from home. (What I find amazing is how they can be continually amazed at this sort of thing.) Most have heard too of the alleged six degrees of separation between any two people in this country. (Actually, under reasonable assumptions each of us is connected to everyone else by an average of two links, although we're not likely to know who the two intermediate parties are.) Another popular variant of the notion concerns the number of movie links between film actors, say between Marlon Brando and Christina Ricci or between Kevin Bacon and anyone else. If A and B appeared together in X, and B and C appeared together in Y, then A is linked to C via these two movies.

Although they may not know of Kevin Bacon and his movies, most mathematicians are familiar with Paul Erdös and his theorems. Erdös, a prolific and peripatetic Hungarian mathematician, wrote hundreds of papers in a variety of mathematical areas during his long life. Many of these had co-authors, who are therefore said to have Erdös number 1. Mathematicians who have written a joint paper with someone with Erdös number 1 are said to have Erdos number 2, and so on.

Ideas about such informal networks lead naturally to the network of all networks, the Internet, and to ways to analyze its structure, shape, and "diameter." How, for example, are the Internet's nearly 1 billion web pages connected? What constitutes a good search strategy? How many links does the average web page contain? What is the distribution of document sizes? Are there many with, say, more than 1,000 links? And, perhaps most intriguingly, how many clicks on average does it take to get from one of two randomly selected documents to another?

A couple of years ago, Albert-Laszló Barabasi, a physics professor at Notre Dame, and two associates, Réka Albert and Hawoong Jeong, published results that strongly suggest that the web is growing and that its documents are linking in a rather collective way that accounts for, among other things, the unexpectedly large number of very popular documents. The increasing number of web pages and the "flocking effect" of many pages pointing to the same popular addresses, causing proportionally more pages to do the same thing, is what leads to a power law.

Barabasi, Albert, and Jeong showed that the probability that a document has k links is roughly proportional to $1/k^3$— or inversely proportional to the third power of k. (I've rounded off; the model actually predicts an exponent of 2.9.) This means, for example, that there are approximately one-eighth as many documents with twenty links as there are documents with ten links since $1/20^3$ is one-eighth of $1/10^3$. Thus the number of documents with k links declines quickly as k increases, but nowhere near as quickly as a normal bell-shaped distribution would predict. This is why the power law distribution has a fatter tail (more instances of very large values of k) than does the normal distribution.

The power laws (sometimes called scaling laws, sometimes Pareto laws) that characterize the web also seem to characterize many other complex systems that organize themselves into a state of skittish responsiveness. The physicist Per Bak, who has made an extensive study of them, described in his book

How Nature Works, claims that such $1/k^m$ laws (for various exponents m) are typical of many biological, geological, musical, and economic processes, and that they tend to arise in a wide variety of complex systems. Traffic jams, to cite a different domain and seemingly unrelated dynamic, also seem to obey a power law, with jams involving k cars occurring with a probability roughly proportional to $1/k^m$ for an appropriate m.

There is even a power law in linguistics. In English, for example, the word "the" appears most frequently and is said to have rank order 1; the words "of," "and," and "to" rank 2, 3, and 4, respectively. "Chrysanthemum" has a much higher rank order. Zipf's Law relates the frequency of a word to its rank order k and states that a word's frequency in a written text is proportional to $1/k^1$; that is, inversely proportional to the first power of k. (Again, I've rounded off; the power of k is close to, but not exactly 1.) Thus a relatively unusual word whose rank order is 10,000 will still appear with a frequency proportional to 1/10,000, rather than essentially not at all as would be the case if word frequencies were described by the tail of a normal distribution. The size of cities also follows a power law with k close to 1, the kth largest city having a population proportional to $1/k$.

One of the most intriguing consequences of the Barabasi-Albert-Jeong model is that because of the power law distribution of links to and from documents on web sites (the nodes of the network), the diameter of the web is only nineteen clicks. By this they mean that you can travel from one arbitrarily selected web page to any other in approximately nineteen clicks, far fewer than had been conjectured. On the other hand, comparing nineteen with the much smaller number of links between arbitrarily selected people, we may wonder why the diameter is as big as it is. The answer is that the average web page contains only seven links, whereas the average person knows hundreds of people.

Even though the web is expected to grow by a power of 10 over the next few years, its diameter will likely grow by only a couple of clicks, from nineteen to twenty-one. The growth and preferential linking assumptions above indicate that the web's diameter D is governed by a logarithmic law; D is a bit more than 2 log(N), where N is the number of documents, presently about 1 billion.

If the Barabasi model is valid (and more work needs to be done), the web is not as unmanageable and untraversable as it often seems. Its documents are much more closely interconnected than they would be if the probability that a document has k links were described by a normal distribution.

What is the relevance of power laws, networks, and diameters to extreme price movements? Investors, companies, mutual funds, brokerages, analysts, and media outlets are connected via a large, vaguely defined network, whose nodes exert influence on the nodes to which they're connected. This network is likely to be more tightly connected and to contain more very popular (and hence very influential) nodes than people realize. Most of the time this makes no difference and price movements, resulting from the sum of a myriad of investors' *independent* zigs and zags, are governed by the normal distribution.

But when the volume of trades is very high, the trades are strongly influenced by relatively few popular nodes—mutual funds, for example, or analysts or media outlets—becoming aligned in their sentiments, and this alignment can create extreme price movements. (WCOM often led the Nasdaq in volume during its slide.) That there exist a few very popular, very connected nodes is, I reiterate, a consequence of the fact that a power law and not the normal distribution governs their frequency. A contagious alignment of this handful of very popular, very connected, very influential nodes will occur more frequently than people expect, as will, therefore, extreme price movements.

Other examples suggest that the exponent m in market power laws, $1/k^m$, may be something other than 3, but the point stands. The trading network is sometimes more herd-like and volatile in its behavior than standard pictures of it acknowledge. The crash of 1929, the decline of 1987, and the recent dot-com meltdown should perhaps not be seen as inexplicable aberrations (or as "just deserts") but as natural consequences of network dynamics.

Clearly much work remains to be done to understand why power laws are so pervasive. What is needed, I think, is something like the central limit theorem in statistics, which explains why the normal curve arises in so many different contexts. Power laws provide an explanation, albeit not an airtight one, for the frequency of bubbles and crashes and the so-called volatility clustering that seem to characterize real markets. They also reinforce the impression that the market is a different sort of beast than that usually studied by social scientists or, perhaps, that social scientists have been studying these beasts in the wrong way.

I should note that my interest in networks and connectivity is not unrelated to my initial interest in WorldCom, which owned not only MCI, but, as I've mentioned twice already, UUNet, "the backbone of the Internet." Obsessions fade slowly.

Economic Disparities and Media Disproportions

WorldCom may have been based in Mississippi, but Bernie Ebbers, who affected an unpretentious, down-home style, wielded political and economic influence foreign to the average Mississippian and the average WorldCom employee. For this he may serve as a synecdoche for the following.

More than a mathematical pun suggests that power laws may have relevance to economic, media, and political power as well as to the stock market. Along various social dimensions, the dynamics underlying power laws might allow for the development of more centers of concentration than we might otherwise expect. This might lead to larger, more powerful economic, media, and political elites and consequent great disparities. Whether or not this is the case, and whether or not great disparities are necessary for complex societies to function, such disparities certainly reign in modern America. Relatively few people, for example, own a hugely disproportionate share of the wealth, and relatively few people attract a hugely disproportionate share of media attention.

The United Nations issued a report a couple of years ago saying that the net worth of the three richest families in the world—the Gates family, the sultan of Brunei, and the Walton family—was greater than the combined gross domestic product of the forty-three poorest nations on Earth. The U.N. statement is misleading in an apples-and-oranges sort of way, but despite the periodic additions, subtractions, and reshufflings of the Forbes 400 and the fortunes of underdeveloped countries, some appropriately modified conclusion no doubt still holds.

(On the other hand, the distribution of wealth in *some* of the poorest nations—where almost everybody is poverty-stricken—is no doubt more uniform than it is here, indicating that relative equality is no solution to the problem of poverty. I suspect that significant, but not outrageous, disparities of wealth are probably more conducive to wealth creation than is relative uniformity, provided the society meets some minimal conditions: It's based on law, offers some educational and other opportunities, and allows for a modicum of private property.)

The dynamic whereby the rich get richer is nowhere more apparent than in the pharmaceutical industry, in which companies understandably spend far, far more money researching lifestyle drugs for the affluent than life-saving drugs for the hundreds of millions of the world's poor people. Instead of trying to come up with treatments for malaria, diarrhea, tuberculosis, and acute lower-respiratory diseases, resources go into treatments for wrinkles, impotence, baldness, and obesity.

Surveys indicate that the ratio of the remuneration of a U.S. firm's CEO to that of the average employee of the firm is at an all-time high of around 500, whether the CEO has improved the fortunes of the company or not, and whether he or she is under indictment or not. (If we assume 250 workdays per year, arithmetic tells us the CEO needs only half a day to make what the employee takes all year to earn.) Professor Edward Wolff of New York University has estimated that the richest 1 percent of Americans own half of all stocks, bonds, and other assets. And Cornell University's Robert Frank has described the spread of the winner-take-all model of compensation from the sports and entertainment worlds to many other domains of American life.

Nero-like arrogance often accompanies such exorbitant compensation. High-tech WorldCom faced a host of problems before its 2002 collapse. Did Bernie Ebbers utilize the company's horde of top-flight technical people (at least the ones who hadn't quit or been fired) to devise a clever strategy to extricate the company from its troubles? No, he cut out free coffee for employees to save money. As Tyco spiraled downward, its CEO, Dennis Kozlowski, spent millions of company dollars on personal items, including a $6,000 shower curtain, a $15,000 umbrella stand, and a $7 million Manhattan apartment.

(Even successful CEOs are not always gentlemen. Oracle's Larry Ellison, a fierce foe of Bill Gates, a couple of years ago admitted to spying on Microsoft. Amusingly, Oracle's sophisticated snoops didn't employ state-of-the-art electronics, but tried to buy the garbage of a pro-Microsoft group in order to examine its contents for clues about Microsoft's public relations plans. I'm talking real cookies here, not the type that Internet sites leave on your computer; scribbled memos and addresses on torn envelopes, not emails and Internet routing numbers; germs and bacteria, not computer viruses.)

What should we make of such stories? Communism, happily, has been discredited, but unregulated and minimally regulated free markets (as evidenced by the behavior of some accountants, analysts, CEOs, and, yes, greedy, deluded, and short-sighted investors) have some obvious drawbacks. Some of the reforms proposed by Congress in 2002 promise to be helpful in this regard, but I wish here only to express disquiet at such enormous and growing economic disparities.

The same steep hierarchy and disproportion that characterize our economic condition affect our media as well. The famous get ever more famous, celebrities become ever more celebrated. (Pick your favorite ten examples here.) Magazines and television increasingly run features asking who's hot and who's not. Even the search engine Google has a version in which surfers can check the topics and people attracting the most hits the previous week. The up-and-down movements of celebrity seem to constitute a kind of market in which almost all the "traders" are technical traders trying to guess what everyone else thinks, rather than value traders looking for worth.

The pattern holds in the political realm as well. In general, on the front page and in the first section of a newspaper, the number one newsmaker is undoubtedly the president of the United States. Other big newsmakers are presidential candidates, members of Congress, and other federal officials.

Twenty years ago, Herbert Gans wrote in *Deciding What's News* that 80 percent of the domestic news stories on television network news concerned these four classes of people; most of the remaining 20 percent covered the other 280 million of us. Fewer than 10 percent of all stories were about abstractions, objects, or systems. Things haven't changed much since then (except on the cable networks where disaster stories, show trials, and terrorist obsessions dominate). Newspapers generally have broader coverage, although studies have found that up to 50 percent of the sources for national stories on the front pages of the *New York Times* and the *Washington Post* were officials of the U.S. government. The Internet has still broader scope, although there, too, one notes strong and unmistakable signs of increasing hierarchy and concentration.

And what about foreign coverage? The frequency of reporting on overseas newsmakers demonstrates the same biases. We hear from heads of state, from leaders of opposition parties or forces, and occasionally from others. The masses of ordinary people are seldom a presence at all. The journalistic rule of thumb that one American equals 10 Englishmen equals 1,000 Chileans equals 10,000 Rwandans varies with time and circumstance, but it does contain an undeniable truth. Americans, like everybody else, care much less about some parts of the world than others. Even the terrorist attack in Bali didn't rate much coverage here, and many regions have no correspondents at all, rendering them effectively invisible.

Such disparities may be a natural consequence of complex societies. This doesn't mean that they need be as extreme as they are or that they're always to be welcomed. It may be that the stock market's recent volatility surges are a leading indicator for even greater social disparities to come.

9 | From Paradox to Complexity

Groucho Marx vowed that he'd never join a club that would be willing to accept him as a member. Epimenides the Cretan exclaimed (almost) inconsistently, "All Cretans are liars." The prosecutor booms, "You must answer Yes or No. Will your next word be 'No'?" The talk show guest laments that her brother is an only child. The author of an investment book suggests that we follow the tens of thousands of his readers who have gone against the crowd.

Warped perhaps by my study of mathematical logic and its emphasis on paradoxes and self-reference, I'm naturally interested in the paradoxical and self-referential aspects of the market, particularly of the Efficient Market Hypothesis. Can it be proved? Can it be disproved? These questions beg a deeper question. The Efficient Market Hypothesis is, I think, neither necessarily true nor necessarily false.

The Paradoxical Efficient Market Hypothesis

If a large majority of investors believe in the hypothesis, they would all assume that new information about a stock would quickly be reflected in its price. Specifically, they would affirm that since news almost immediately moves the price up or down, and since news can't be predicted, neither can changes in stock prices. Thus investors who subscribe to the Efficient Market Hypothesis would further believe that looking for trends and analyzing companies' fundamentals is a waste of time. Believing this, they won't pay much attention to new developments. But if relatively few investors are looking for an edge, the market will not respond quickly to new information. In this way an overwhelming belief in the hypothesis ensures its falsity.

To continue with this cerebral somersault, recall now a rule of logic: Sentences of the form "H implies I" are equivalent to those of the form "not I implies not H." For example, the sentence "heavy rain implies that the ground will be wet" is logically equivalent to "dry ground implies the absence of heavy rain." Using this equivalence, we can restate the claim that overwhelming belief in the Efficient Market Hypothesis leads to (or implies) its falsity. Alternatively phrased, the claim is that if the Efficient Market Hypothesis is true, then it's not the case that most investors believe it to be true. That is, if it's true, most investors believe it to be false (assuming almost all investors have an opinion and each either believes it or disbelieves it).

Consider now the inelegantly named Sluggish Market Hypothesis, the belief that the market is quite slow in responding to new information. If the vast majority of investors believe the Sluggish Market Hypothesis, then they all would believe that looking for trends and analyzing companies is well worth their time and, by so exercising themselves, they would bring about an efficient market. Thus, if most investors believe the Sluggish Market Hypothesis is true, they will by their actions make the Efficient Market Hypothesis true. We conclude that if the Efficient Market Hypothesis is false, then it's not the case that most investors believe the

Sluggish Market Hypothesis to be true. That is, if the Efficient Market Hypothesis is false, then most investors believe it (the EMH) to be true. (You may want to read over the last few sentences in a quiet corner.)

In summary, if the Efficient Market Hypothesis is true, most investors won't believe it, and if it's false, most investors will believe it. Alternatively stated, the Efficient Market Hypothesis is true if and only if a majority believes it to be false. (Note that the same holds for the Sluggish Market Hypothesis.) These are strange hypotheses indeed!

Of course, I've made some big assumptions that may not hold. One is that if an investor believes in one of the two hypotheses, then he disbelieves in the other, and almost all believe in one or the other. I've also assumed that it's clear what "large majority" means, and I've ignored the fact that it sometimes requires very few investors to move the market. (The whole argument could be relativized to the set of knowledgeable traders only.)

Another gap in the argument is that any suspected deviations from the Efficient Market Hypothesis can always be attributed to mistakes in asset pricing models, and thus the hypothesis can't be conclusively rejected for this reason either. Maybe some stocks or kinds of stock are riskier than our pricing models allow for and that's why their returns are higher. Nevertheless, I think the point remains: The truth or falsity of the Efficient Market Hypothesis is not immutable but depends critically on the beliefs of investors. Furthermore, as the percentage of investors who believe in the hypothesis itself varies, the truth of the hypothesis varies inversely with it.

On the whole, most investors, professionals on Wall Street, and amateurs everywhere, disbelieve in it, so for this reason I think it holds, but only approximately and only most of the time.

The Prisoner's Dilemma and the Market

So you don't believe in the Efficient Market Hypothesis. Still, it's not enough that you discover simple and effective investing rules. Others must not find out what you're doing, either by inference or by reading your boastful profile in a business magazine. The reason for secrecy, of course, is that without it, simple investing rules lead to more and more complicated ones, which eventually lead to zero excess returns and a reliance on chance.

This inexorable march toward increased complexity arises from the actions of your co-investors, who, if they notice (or infer, or are told) that you are performing successfully on the basis of some simple technical trading rule, will try to do the same. To take account of their response, you must complicate your rule and likely decrease your excess returns. Your more complicated rule will, of course, also inspire others to try to follow it, leading to further complications and a further decline in excess returns. Soon enough your rule assumes a near-random complexity, your excess returns are reduced essentially to zero, and you're back to relying on chance.

Of course, your behavior will be the same if you learn of someone else's successful performance. In fact, a situation arises that is clarified by the classic "prisoner's dilemma," a useful puzzle originally framed in terms of two people in prison.

Suspected of committing a major crime, the two are apprehended in the course of committing some minor offense. They're then interrogated separately, and each is given the choice of confessing to the major crime and thereby implicating his partner or remaining silent. If they both remain silent, they'll each get one year in prison. If one confesses and the other doesn't, the one who confesses will be rewarded by being set free, while the other one will get a five-year term. If they both confess, they can both expect to spend three years in prison. The cooperative option (cooperative with the other prisoner, that is) is to remain silent, while the non-cooperative option is to confess. Given the payoffs and human psychology, the most likely outcome is for both to confess; the best outcome for the pair *as a pair* is for both to remain silent; the best outcome for each prisoner *as an individual* is to confess and have one's partner remain silent.

The charm of the dilemma has nothing to do with any interest one might have in prisoners' rights. (In fact, it has about as much relevance to criminal justice as the four-color-map theorem has to geography.) Rather, it provides the logical skeleton for many situations we face in everyday life. Whether we're negotiators in business, spouses in a marriage, or nations in a dispute, our choices can often be phrased in terms of the prisoner's dilemma. If both (all) parties pursue their own interests exclusively and do not cooperate, the outcome is worse for both (all) of them; yet in any given situation, any given party is better off not cooperating. Adam Smith's invisible hand ensuring that individual pursuits bring about group well-being is, at least in these situations (and some others), quite arthritic.

The dilemma has the following multi-person market version: Investors who notice some exploitable stock market anomaly may either act on it, thereby diminishing its effectiveness (the non-cooperative option) or ignore it, thereby saving themselves the trouble of keeping up with developments (the cooperative option). If some ignore it and others act on it, the latter will receive the biggest payoffs, the former the smallest. As in the standard prisoner's dilemma, the logical response for any player is to take the non-cooperative option and act on any anomaly likely to give one an edge. This response leads to the "arms race" of ever more complex technical trading strategies. People search for special knowledge, the result eventually becomes common knowledge, and the dynamic between the two generates the market.

This searching for an edge brings us to the social value of stock analysts and investment professionals. Although the recipients of an abundance of bad publicity in recent years, they provide a most important service: By their actions, they help turn special knowledge into common knowledge and in the process help make the market relatively efficient. Absent a draconian rewiring of human psychology and an accompanying draconian rewiring of our economic system, this accomplishment is an impressive and vital one. If it means being "noncooperative" with other investors, then so be it. Cooperation is, of course, generally desirable, but cooperative decisionmaking among investors seems to smack of totalitarianism.

Pushing the Complexity Horizon

The complexity of trading rules admits of degrees. Most of the rules to which people subscribe are simple, involving support levels, P/E ratios, or hemlines and Super Bowls, for example. Others, however, are quite convoluted and conditional. Because of the variety of possible rules, I want to take an oblique and abstract approach here. The hope is that this approach will yield insights that a more pedestrian approach misses. Its key ingredient is the formal definition of (a type of) complexity. An intuitive understanding of this notion tells us that someone who remembers his eight-digit password by means of an elaborate, long-winded saga of friends' addresses, children's ages, and special anniversaries is doing something silly. Mnemonic rules make sense only when they're shorter than what is to be remembered.

Let's back up a bit and consider how we might describe the following sequences to an acquaintance who couldn't see them. We may imagine the 1s to represent upticks in the price of a stock and the 0s downticks or perhaps up-and-down days.

1. 0 1 0 1 0 1 0 1 0 1 0 1 0 1 0 1 0 1 0 1 0 1 0 1 0 ...
2. 0 1 0 1 1 0 1 0 1 0 1 0 1 1 0 1 0 1 0 1 0 1 0 1 1 ...
3. 1 0 0 0 1 0 1 1 0 1 1 0 1 1 0 0 0 1 0 1 0 1 1 0 0 ...

The first sequence is the simplest, an alternation of 0s and 1s. The second sequence has some regularity to it, a single 0 alternating sometimes with a 1, sometimes with two 1s, while the third sequence doesn't seem to manifest any pattern at all. Observe that the precise meaning of " ... " in the first sequence is clear; it is less so in the second sequence, and not at all clear in the third. Despite this, let's assume that these sequences are each a trillion bits long (a bit is a 0 or a 1) and continue on "in the same way."

Motivated by examples like this, the American computer scientist Gregory Chaitin and the Russian mathematician A. N. Kolmogorov defined the complexity of a sequence of 0s and 1s to be the length of the shortest computer program that will generate (that is, print out) the sequence in question.

A program that prints out the first sequence above can consist simply of the following recipe: print a 0, then a 1, and repeat a half trillion times. Such a program is quite short, especially compared to the long sequence it generates. The complexity of this first trillion-bit sequence may be only a few hundred bits, depending to some extent on the computer language used to write the program.

A program that generates the second sequence would be a translation of the following: Print a 0 followed by either a single 1 or two 1s, the pattern of the intervening 1s being one, two, one, one, one, two, one, one, and so on. Any program that prints out this trillion-bit sequence would have to be quite long so as to fully specify the "and so on" pattern of the intervening 1s. Nevertheless, because of the regular alternation of 0s and either one or two 1s, the shortest such program will be considerably shorter than the trillion-bit sequence it generates. Thus the complexity of this second sequence might be only, say, a quarter trillion bits.

With the third sequence (the commonest type) the situation is different. This sequence, let us assume, remains so disorderly throughout its trillion-bit length that no program we might use to generate it would be any shorter than the sequence itself. It never repeats, never exhibits a pattern. All any program can do in this case is dumbly list the bits in the sequence: print 1, then 0, then 0, then 0, then 1, then 0, then 1, ... There is no way the ... can be compressed or the program shortened. Such a program will be as long as the sequence it's supposed to print out, and thus the third sequence has a complexity of approximately a trillion.

A sequence like the third one, which requires a program as long as itself to be generated, is said to be random. Random sequences manifest no regularity or order, and the programs that print them out can do nothing more than direct that they be copied: print 1 0 0 0 1 0 1 1 0 1 1 .... These programs cannot be abbreviated; the complexity of the sequences they generate is equal to the length of these sequences. By contrast, ordered, regular sequences like the first can be generated by very short programs and have complexity much less than their length.

Returning to stocks, different market theorists will have different ideas about the likely pattern of 0s and 1s (downs and upticks) that can be expected. Strict random walk theorists are likely to believe that sequences like the third characterize price movements and that the market's movements are therefore beyond the "complexity horizon" of human forecasters (more complex than we, or our brains, are, were we expressed as sequences of 0s and 1s). Technical and fundamental analysts might be more inclined to believe that sequences like the second characterize the market and that there are pockets of order amidst the noise. It's hard to imagine anyone believing that price movements follow sequences as regular as the first except, possibly, those who send away "only $99.95 for a complete set of tapes that explain this revolutionary system."

I reiterate that this approach to stock price movements is rather stark, but it does nevertheless "locate" the debate. People who believe there is some pattern to the market, whether exploitable or not, will believe that its movements are characterized by sequences of complexity somewhere between those of type two and type three above.

A rough paraphrase of Kurt Godel's famous incompleteness theorem of mathematical logic, due to the aforementioned Gregory Chaitin, provides an interesting sidelight on this issue. It states that if the market were random, we might not be able to prove it. The reason: encoded as a sequence of 0s and 1s, a random market would, it seems plausible to assume, have complexity greater than that of our own were we also so encoded; it would be beyond our complexity horizon. From the definition of complexity it follows that a sequence can't generate another sequence of greater complexity than itself. Thus if a person were to predict the random market's exact gyrations, the market would have to be less complex than the person, contrary to assumption. Even if the market isn't random, there remains the possibility that its regularities are so complex as to be beyond our complexity horizons.

In any case, there is no reason why the complexity of price movements as well as the complexity of investor/computer blends cannot change over time. The more inefficient the market is, the smaller the complexity of its price movements, and the more likely it is that tools from technical and fundamental analysis will prove useful. Conversely, the more efficient the market is, the greater the complexity of price movements, and the closer the approach to a completely random sequence of price changes.

Outperforming the market requires that one remain on the cusp of our collective complexity horizon. It requires faster machines, better data, improved models, and the smarter use of mathematical tools, from conventional statistics to neural nets (computerized learning networks, the connections between the various nodes of which are strengthened or weakened over a period of training). If this is possible for anyone or any group to achieve, it's not likely to remain so for long.

Game Theory and
Supernatural Investor/Psychologists

But what if, contrary to fact, there were an entity possessing sufficient complexity and speed that it was able with reasonably high probability to predict the market and the behavior of individuals within it? The mere existence of such an entity leads to Newcombe's paradox, a puzzle that calls into question basic principles of game theory.

My particular variation of Newcombe's paradox involves the World Class Options Market Maker (WCOMM), which (who?) claims to have the power to predict with some accuracy which of two alternatives a person will choose. Imagine further that WCOMM sets up a long booth on Wall Street to demonstrate its abilities.

WCOMM explains that it tests people by employing two portfolios. Portfolio A contains a $1,000 treasury bill, whereas portfolio B (consisting of either calls or puts on WCOM stock) is either worth nothing or $1,000,000. For *each* person in line at the demonstration, WCOMM has reserved a portfolio of each type at the booth and offers each person the following choice: He or she can choose to take portfolio B *alone* or choose to take *both* portfolios A and B. However, and this is crucial, WCOMM also states that it has used its unfathomable powers to analyze the psychology, investment history, and trading style of everyone in line as well as general market conditions, and if it believes that a person will take both portfolios, it has ensured that portfolio B will be worthless. On the other hand, if WCOMM believes that a person will trust its wisdom and take only portfolio B, it has ensured that portfolio B will be worth $1,000,000. After making these announcements, WCOMM leaves in a swirl of digits and stock symbols, and the demonstration proceeds.

Investors on Wall Street see for themselves that when a person in the long line chooses to take both portfolios, most of the time (say with probability 90 percent) portfolio B is worthless and the person gets only the $1,000 treasury bill in portfolio A. They also note that when a person chooses to take the contents of portfolio B alone, most of the time it's worth $1,000,000.

After watching the portfolios placed before the people in line ahead of me and seeing their choices and the consequences, I'm finally presented with the two portfolios prepared for me by WCOMM. Despite the evidence I've seen, I see no reason not to take both portfolios. WCOMM is gone, perhaps to the financial district of London or Frankfurt or Tokyo, to make similar offers to other investors, and portfolio B is either worth $1,000,000 or not, so why not take both portfolios and possibly get $1,001,000. Alas, WCOMM read correctly the skeptical smirk on my face and after opening my portfolios, I walk away with only $1,000. My portfolio B contains call options on WCOM with a strike price of 20, when the stock itself is selling at $1.13.

The paradox, due to the physicist William Newcombe (not the Newcomb of Benford's Law, but the same mocking four

letters WCOM) and made well-known by the philosopher
Robert Nozick, raises other issues. As mentioned, it makes
problematical which of two game-theoretic principles one
should use in making decisions, principles that shouldn't conflict.

The "dominance" principle tells us to take both portfolios
since, whether portfolio B contains options worth $1,000,000
or not, the value of two portfolios is at least as great as the
value of one. (If portfolio B is worthless, $1,000 is greater
than $0, and if portfolio B is worth $1,000,000, $1,001,000
is greater than $1,000,000.)

On the other hand, the "maximization of expected value"
principle tells us to take only portfolio B since the expected
value of doing so is greater. (Since WCOMM is right about 90
percent of the time, the expected value of taking only portfolio
B is (.90 × $1,000,000) + (.10 × $0), or $900,000, whereas the
expected value of taking both is (.10 × $1,001,000) + (.90 ×
$1,000), or $101,000.) The paradox is that both principles
seem reasonable, yet they counsel different choices.

This raises other general philosophical matters as well, but
it reminds me of my resistance to following the WCOM-
fleeing crowd, most of whose B portfolios contained puts on
the stock worth $1,000,000.

One conclusion that seems to follow from the above is that
such supernatural investor/psychologists are an impossibility.
For better or worse, we're on our own.

Absurd Emails and the WorldCom Denouement

A natural reaction to the vagaries of chance is an attempt at
control, which brings me to emails regarding WorldCom that,
Herzog-like, I sent to various influential people. I had grown
tired of carrying on one-sided arguments with CNBC's always

perky Maria Bartiromo and always apoplectic James Cramer as they delivered the relentlessly bad news about WorldCom. So in fall 2001, five or six months before its final swoon, I contacted a number of online business commentators critical of WorldCom's past performance and future prospects. Having spent too much time in the immoderate atmosphere of WorldCom chatrooms, I excoriated them, though mildly, for their shortsightedness and exhorted them to look at the company differently.

Finally, out of frustration with the continued decline of WCOM stock, I emailed Bernie Ebbers, then the CEO, in early February 2002 suggesting that the company was not effectively stating its case and quixotically offering to help by writing copy. I said I'd invested heavily in WorldCom, as did family and friends at my suggestion, that I could be a persuasive wordsmith when I believed in something, and WorldCom, I believed, was well positioned but dreadfully undervalued. UUNet, the "backbone" of much of the Internet, was, I fatuously informed the CEO of the company, a gem in and of itself.

I knew, even as I was writing them, that sending these electronic epistles was absurd, but it gave me the temporary illusion of doing something about this recalcitrant stock other than dumping it. Investing in it had originally seemed like a no-brainer. The realization that doing so had indeed been a no-brainer was glacially slow in arriving. During the 2001–2002 academic year, I took the train once a week from Philadelphia to New York to teach a course on "numbers in the news" at the Columbia School of Journalism. Spending the two and a half hours of the commute out of contact with WCOM's volatile movements was torturous, and upon emerging from the subway, I'd run to my office computer to check what had happened. Not exactly the behavior of a sage long-term investor; my conduct even then suggested to me a rather dim-witted addict.

Recalling the two or three times I almost got out of the stock is dispiriting as well. The last time was in April 2002. Amazingly, I was even then still somewhat in thrall to the idea of averaging down, and when the price dipped below $5, I bought more WCOM shares. Around the middle of the month, however, I did firmly and definitively resolve to sell. By Friday, April 19, WCOM had risen to over $7, which would have allowed me to recoup at least a small portion of my losses, but I didn't have time to sell that morning. I had to drive to northern New Jersey to give a long-promised lecture at a college there. When it was over, I wondered whether to return home to sell my shares or simply use the college's computer to log onto my Schwab account to do so. I decided to go home, but there was so much traffic on the cursed New Jersey turnpike that afternoon that I didn't arrive until 4:05, after the market had closed. I had to wait until Monday.

Investors are often nervous about holding volatile stocks over the weekend, and I was no exception. My anxiety was well-founded. Later that evening there was news about impending cuts in WorldCom's bond ratings and another announcement from the SEC regarding its comprehensive investigation of the company. The stock lost more than a third of its value by Monday, when I did finally sell the stock at a huge loss. A few months later the stock completely collapsed to $.09 upon revelations of massive accounting fraud.

Why had I violated the most basic of investing fundamentals: Don't succumb to hype and vaporous enthusiasm; even if you do, don't put too many eggs in one basket (especially with the uncritical sunny-side up); even if you do this, don't forget to insure against sudden drops (say with puts, not calls); and even if you do this too, don't buy on margin. After selling my shares, I felt as if I were gradually and groggily coming out of a self-induced trance. I'd long known about one of the earliest "stock" hysterias on record, the seventeenth century tulip bulb craze in Holland. After its collapse, people also spoke of waking up and realizing that they were stuck with nearly worthless bulbs and truly worthless options to buy more of them. I smiled ruefully at my previous smug dismissal of people like the tulip bulb "investors." I was as vulnerable to transient delirium as the dimmest bulb-buying bulbs.

I've followed the ongoing drama of the WorldCom story— the fraud investigations, various prosecutions, new managers, promised reforms, and court settlements—and, oddly perhaps, the publicity surrounding the scandals and their aftermath has distanced me from my experience and lessened its intensity. My losses have become less a small personal story and more (a part of) a big news story, less a result of my mistakes and more a consequence of the company's behavior. This shifting of responsibility is neither welcome nor warranted. For reasons of fact and of temperament, I continue to think of myself as having been temporarily infatuated rather than deeply victimized. Remnants of my fixation persist, and I still sometimes wonder what might have happened if WorldCom's deal with Sprint hadn't been foiled, if Ebbers hadn't borrowed $400 million (or more), if Enron hadn't imploded, if this or that or the other event hadn't occurred before I sold my shares. My recklessness might then have been seen as daring. Post hoc stories always seem right, whatever the pre-existing probabilities.

One fact remains incontrovertible: Narratives and numbers coexist uneasily on Wall Street. Markets, like people, are largely rational beasts occasionally provoked and disturbed by their underlying animal spirits. The mathematics discussed in this book is often helpful in understanding (albeit not beating) the market, but I'd like to end with a psychological caveat. The basis for the application of the mathematical tools discussed herein is the sometimes shifty and always shifting attitudes of investors. Since these psychological states are to a large extent imponderable, anything that depends on them is less exact than it appears.

The situation reminds me a bit of the apocryphal story of the way cows were weighed in the Old West. First the cowboys would find a long, thick plank and place the middle of it on a large, high rock. Then they'd attach the cow to one end of the plank with ropes and tie a large boulder to the other end. They'd carefully measure the distance from the cow to the rock and from the boulder to the rock. If the plank didn't balance, they'd try another big boulder and measure again. They'd keep this up until a boulder exactly balanced the cattle. After solving the resulting equation that expresses the cow's weight in terms of the distances and the weight of the boulder, there would be only one thing left for them to do: They would have to guess the weight of the boulder. Once again the mathematics may be exact, but the judgments, guesses, and estimates supporting its applications are anything but.

More apropos of the self-referential nature of the market would be a version in which the cowboys had to guess the weight of the cow whose weight varied depending on their collective guesses, hopes, and fears. Bringing us full circle to Keynes's beauty contest, albeit in a rather forced, more bovine mode, I conclude that despite rancid beasts like WorldCom, I'm still rather fond of the pageant that is the market. I just wish I had a better (and secret) method for weighing the cows.

Bibliography

The common ground and intersection of mathematics, psychology, and the market is a peculiar interdisciplinary niche (even without the admixture of memoir). There are within it many mathematical tomes and theories ostensibly relevant to the stock market. Most are not. There are numerous stock-picking techniques and strategies that appear to be very mathematical. Not many are. There are a good number of psychological accounts of trading behavior, and a much smaller number of mathematical approaches to psychology, but much remains to be discovered. A few pointers to this inchoate, nebulously defined, yet fascinating area follow:

Bak, Per, *How Nature Works*, New York, Springer-Verlag, 1996.

Barabasi, Albert-Laszlo, *Linked: The New Science of Networks*, New York, Basic Books, 2002.

Dodd, David L., and Benjamin Graham, *Security Analysis*, New York, McGraw-Hill, 1934.

Fama, Eugene F., "Efficient Capital Markets, II," *Journal of Finance*, December 1991.

Gilovich, Thomas, *How We Know What Isn't So*, New York, Simon and Schuster, 1991.

Hart, Sergiu, and Yair Tauman, "Market Crashes Without Exogenous Shocks," The Hebrew University of Jerusalem, Center for Rationality DP–124, December 1996 (forthcoming in *Journal of Business*).

Kritzman, Mark P., *Puzzles of Finance,* New York, John Wiley, 2000.

Lefevre, Edwin, *Reminiscences of a Stock Operator,* New York, John Wiley, 1994 (orig. 1923).

Lo, Andrew, and Craig MacKinlay, *A Non-Random Walk Down Wall Street*, Princeton, Princeton University Press, 1999.

Malkiel, Burton, *A Random Walk Down Wall Street*, New York, W. W. Norton, 1999 (orig. 1973).

Mandelbrot, Benoit, "A Multifractal Walk Down Wall Street," *Scientific American*, February 1999.

Paulos, John Allen, *Once Upon a Number*, New York, Basic Books, 1998.

Ross, Sheldon, *Probability*, New York, Macmillan, 1976.

Ross, Sheldon, *Mathematical Finance,* Cambridge, Cambridge University Press, 1999.

Siegel, Jeremy J., *Stocks for the Long Run*, New York, McGraw-Hill, 1998.

Shiller, Robert J., *Irrational Exuberance*, Princeton, Princeton University Press, 2000.

Taleb, Nassim Nicholas, *Fooled by Randomness*, New York, Texere, 2001.

Thaler, Richard, *The Winner's Curse*, Princeton, Princeton University Press, 1992.

Tversky, Amos, Daniel Kahneman, and Paul Slovic, *Judgment Under Uncertainty: Heuristics and Biases,* Cambridge, Cambridge University Press, 1982.

Index